Basic Common Sense

Further Radical Ecological
Thought Experiments

Mozart & Reason Wolfe books

Urania Press
One Earth Many Worlds, by Alan Wittbecker
RE: viewing thinking & turning
Good Forestry from Good Theories & Good Practices
(O)utopias or (E)utopias, by Alan Wittbecker
Topopoetics, by Alan Wittbecker
Global Emergency Actions, by Alan Wittbecker
Eutopias: Making Good Places Ecologically & Culturally
Redesigning the Planet: Foundations
Redesigning the Planet: Local Systems
Redesigning the Planet: Regions
Redesigning the Planet: Global Ecological Design

Calliope Press
Two Diaries, by Marcus Rian
Amphibian Dreams, by A. M. Caratheodory
Wild Apples, by A. M. Caratheodory
Light from a Vanished Forest, by A. M. Caratheodory
Carbon Dreams, by Violet Reason
Tropomorphoses, by Yulalona Lopez
Night Wolves, by Yulalona Lopez
Coyote Redivivus (11-book series), by Yulalona Lopez
Coyote Redux, by violet reason & Yulalona Lopez
Coyote Recharged, by Yulalona Lopez

Clio Press
The Thesis, by Marcus Rian
Waiting for Better Times in Bulgaria, by Conor Ciaran
Guarded by Trees, by A. M. Caratheodory
Musings, ed. by Crawford Washington
Murmur, ed. by C. A. Washington
Shadow Masks, ed. by Crawford Washington
Poetic Archaeology of the Flesh
Global Government

Basic
Common
Sense

Further
Radical Ecological
Thought Experiments

On Making Effective Changes
During a Global Emergency
Using Ecological Perspectives
And Coordinated Cultural Actions

Alan Wittbecker

Urania Science Press
Mozart & Reason Wolfe, Ltd
Sarasota 2016

Cover Design: 2016, Rian Garcia Calusa
Book Design by Rian Garcia Calusa, Sarasota, Florida
Rian Garcia Calusa: www.riangarciacalusa.com, designs@ ...
Other graphics and photographs: Alan Wittbecker

Published by Mozart & Reason Wolfe Ltd, Urania Science Press
for SynGeo ArchiGraph
    Mail: Post Office Box 370, Tallevast, Florida 34270
    Email: Design@SynGeo.org, Mozart@ReasonWolfe.com

For more information on sites and projects:
    SynGeo ArchiGraph LLC: www.syngeo.org
    Ecoforestry Institute: www.ecoforestry.net
    G. P. Marsh Institute: www.marshinstitute.net
    Eutopian Ecologists: www.eutopias.net

Publisher's Cataloging-in-Publication Data
Alan Wittbecker
Basic Common Sense / Alan Wittbecker
Includes bibliographical references and index.

ISBN-13: 978-0911385687
ISBN-10: 0911385681 (softcover)
1. Common Sense. 2. Thought Experiments.
3. Ecological Design. 4. Ecosystem ecology. 5. Deep Ecology.
I. Title. GF75.W64 2016

Printed in the United States of America
First Edition

1 2 3 4 5 6 7 8 9 10

Contents

Acknowledgments

To Dr Michael W. Fox for continuing the conversation, for challenging me to continue through the questions, and finally for providing examples of good science writing in his passionate output of books on animal science and health. For Deanna Krantz for other conversations on animal welfare and health relating to IPAN's work establishing an animal hospital for elephants, cows, donkeys, dogs, and for any animal in need, in India.

To John B. Cobb Jr for patiently leading me through the 'tangled bank' of my errors on process thinking and ecological economics. I am grateful that he continues to encourage me in areas he has pioneered in theology, economics, ecology, and city design. To Paolo Soleri for graciously hosting me at Arcosanti to work on early portions of the book. He also provided texts, graphics and assistance for an upcoming book on *Arcologies*. To Arne Naess for discussions on peace and deep ecology, between climbing and boxing practice. I have to admit that the cold mountain air in Norway was conducive to deep thinking about ecology. He arranged for me to give a series of departmental and university lectures on population and wilderness reservation one semester as his post-doc. To Garrett Hardin for his correspondence on topics related to global things, populations and commons (not to mention the incredibly ugly Darwinian postcards we used to exchange ideas). To Buckminster Fuller for rejecting my untenable ideas on technology and listening to my tentative suggestions for him to pay more attention to ecological and political perspectives. To Paul Shepard for refining the ideas of cultural madness and for inviting me to help build his house. I was finishing my own at the time.

To Thomas Berry for sending me drafts of his work on approaching the earth and for reading my draft articles; it was an enjoyable exchange of handwritten papers. To Eugene Odum for sharing thoughts and time on ecological science and perspectives during our meetings at conferences on ecology. To Alan Drengson for helping me on the path to an ecoforestry approach to forest management and restoration, and for helping with the grants, ecoforestry books and conference invitations (e.g., *50 Wildest Thinkers of the 22nd Century*). To Leo Bustad, for encouraging me to teach a veterinary ethics course, knowing that I could not finish my DVM.

To Craig Dillard, for decades of wild conversations and great dinners at our hand-built, stained-glass restaurant, The Seasons (he and Linda Dillard were the driving forces behind its success, 1984-1999). To Devorah Bell, Caroline Hagen, Linda Martin, Margaret Wittbecker, Hanna Metzger, James Luck, Janet Wampler, and Neil Keefe for reading and questioning portions of this work. To Marcella Crider, for her unquestioning support and suggestions to simplify biological and thermodynamic topics.

Preface: Thomas Paine & Common Sense

What better title can a book on revolution have than "*Common Sense*"? And not just on revolution, but also on the ideals of the enlightenment, the equality of humanity, and how it should represent itself for its greatest happiness. And not just that, but by extension on all beings, all life on earth, and the entire planet as someone deserving consideration? Despite the efforts of Thomas Paine and other revolutionaries, the problems of slavery, violence and stupidity have not changed or gone away in an industrial civilization. They still need to rooted out and neutralized.

In café society, reformed churches and local bars in the eighteenth-century, Thomas Paine was a well-educated revolutionary, in touch with his contemporaries, who also sought independence and a new kind of state, not committed to war and desolation like England. Yet, within a generation, the radical ideas of revolution were rendered domestic and safe by patriotism and nostalgia. The strong bonds of the original framers of the constitution—Franklin, Jefferson, Adams, Washington, and the rest—were frayed and broken, and finally ignored by a new generation of thinkers and politicians, who analyzed freedom into a tired tradition, replacing it with the moral self-superiority of money and games through detachment and dissociation.

Paine's utopian vision was recast as the ravings of a drunken extremist. The dream of an independent classic republic, shining with knowledge and virtue, was replaced by the pedestrian goals of materialism and prosperity in a divided state. Simple, comfortable anti-intellectual, anti-historical competition for money and power—that constant idiotic cycling of money into power and back again by the undeserving recipients of inherited wealth. Freedoms of expression, assembly, and education— the radical achievements of the enlightened American Revolution—were routinely dismissed or rerouted if they disturbed the current flow.

The industrial revolution was a bad deal for people. Not only was the pace of work set by machines, but people became their slaves. Or rather interchangeable units that could be oppressed as their independent way of life was annihilated. Even workers' organizations to protect skills were banned and finally marginalized into social eating clubs. We have all been trapped by the comforts and usefulness of the industrial way, by the promise of security in sameness and flatscapes, and by the insistence on obedience. We remain chained by ignorance and superstition.

Paine explained the ideas of tyranny and independence in such a basic manner that any reader, or listener, could understand. The application of the enlightenment ideas of good and equality was *not* some crusty philosophical exercise, it was a necessary action! And, he presented them in a short, inexpensive pamphlet. He had complete freedom of subject and expression; he could be serious or seditious, abusive or enlightening. He could present it as a philosophical lecture or a pulpit sermon. He could exhort to war or outline a model of government. He wrote the pamphlet, then paid the paper and printer's fees to distribute copies of it.

The pamphlet was an attack on everything feudal about Europe, from unstable, destructive hereditary thrones to the demanding privileges of the accidental class. The logical strength of his arguments convinced readers that the English contract had to be dissolved by them. He calmed the readers' fears about isolation and failure by showing how their aspirations would succeed, because a monarch (or tyrant or millionaire now, too) had no power beyond what the citizens granted to her. He heated up the Declaration of Independence, confident that American independence would not destroy the successful English system. He seized the attention of colonists, portraying them as pioneers and creators rather than as traitors or loyalists, with a pamphlet on political self-help, until it could be turned to action to make a better world for coming generations.

And now, so many things addressed in the pamphlet have resurfaced or been reiterated in slightly different forms: The divine right of the wealthy or merely rich to multiply their money without constraints and be treated as superior; the wickedness of government in restraining our vices and in promoting the common good (and supporting it with taxes); the rights of one religion, Christianity, to replace formal laws with its deadly rules on social and private things like abortion or alternate forms of sexuality, and to ban other faiths; the patent crap that humanity is independent of nature and not subject to natural laws, while using or killing all other living beings; and finally, that it is okay to own other people, animals or things as slaves, as long as they tacitly agree by not killing us.

This is why more common sense is necessary: We have *no* moral right to enslave people or drive plants and animals to extinction; we have *no* progressive right to destroy older, smaller cultures unless they embrace industrialism. It is a very *bad idea* to pretend that we are independent from nature while relying on wild ecosystems for services like clean water or support for our food from agriculture; and, it is a *worse idea* to ignore the trillions of dollars of natural capital by touting the foundation of virtual computer worlds.

We *need* a Declaration of *Interdependence* (with living nature and its thriving networks). We *need* another continental congress to rework the charter to secure freedom and representation in a better form; and, we *need* a constitution that can prescribe the actions and interrelationships of all human groups, from sewing clubs to international corporations.

It is the intent of this book on basic common sense to serve as a reminder that the core of enlightenment beliefs—that people need to be free to choose, that people are born with certain inalienable rights, that government is empowered by its citizens, that the nation works by rule of law, that all are equal under law, and that the nation has a duty to help all achieve the best they can—are still valid and need to be restated, to re-inspire people politically and spiritually.

Thomas Paine risked hanging for treason for his writings. This author risks mild insults or painful indifference. Perhaps Paine was too raw and harsh, too unsettling. Perhaps this effort now is not raw enough. Paine sought to open every set of eyes, to see that 'mankind' was *not* born with saddles on their backs, to be ridden for the convenience of a few with boots and spurs. American History was, and still is, a grand experiment for bringing people together peacefully to form a better nation; these small experiments try to continue that tradition. One thing has come full circle—this author has complete control of this 'big pamphlet,' of the subject and its expression, and of the paper and printing fees (5% over cost). The author asks that, more than just giving lip service to our rights, we need to commit our undiminished devotion to them. The author reminds all of you that we, as Thomas Paine said, "have it in our power to begin the world over again." *Can we begin now?*

Sarasota, June 2016

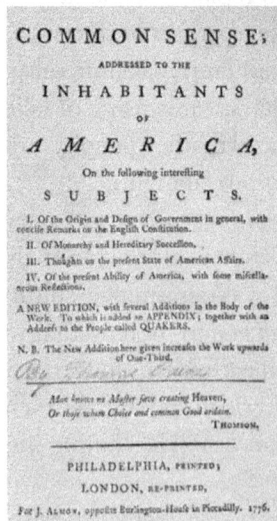

Figure 12. Common Sense pamphlet cover.

Common Sense & Thought Experiments

In science, some experiments turned investigations in new directions. For instance, Albert Einstein started to think about Relativity first as a thought experiment (*Gedankenexperiment* in German): What would happen if the observer were traveling at the speed of light? When he was a schoolboy, he had asked himself whether he could see himself in a mirror if he was riding on a beam of light; the answer is yes, according to an older Einstein, because he and the mirror are in the same local frame of reference (although in thought the observer can be in two frames simultaneously, and things may seem strange from one or other frames). Earlier, Isaac Newton thought that a falling apple was governed by the same laws as the orbit of a planet. These thought experiments in physics demonstrated that thinking in a productive way could be as effective as mixing substances in test tubes or measuring weights or speeds.

Thought experiments can result in a new perspective, which can lead to new knowledge and to better ways of doing things. Of course, the human brain may subconsciously create experiments to improve its ability to make decisions when the environment is uncertain. Conscious thought experiments are useful when observer participation is physically impossible. This was true for the extreme cases of physics and astrophysics. It should be required in cases dealing with large numbers of living beings in a system (ecosystem) or with complex and conscious animals, especially human beings and other mammals. Most people agree that those kinds of experiments are immoral. Yet, it used to be in biology and veterinary medicine that many courses ended in the death of the subjects (and for people like me, failing required courses for doing the 'right' thing by freeing them to a farm). That is happening less often now, as a result of a change in consciousness—what Albert Schweitzer noted is a trend towards the reverence for life—and with some new educational courses (later, I was able to design and teach 'Moral Issues in Science,' 1973 and 'Veterinary Medical Ethics,' 1978). Some cultures, military victors in the 1930s and 40s, had forced crude experimentation on conquered cultures; the knowledge gained was trivial, and these have been universally condemned as violating ethical codes. Other harmful, impossible, large, incomplete, or complex experiments are sometimes tried first, if at all, as thought experiments. The thought experiments presented here are offered as solutions to some complex social or cultural problems.

The Return of Common Sense

Magazines and books are bawling about advances to drive progress, save the planet, and improve human life. Technology is expected to solve problems as well as push progress along. This is not to say that technology has no amazing uses—it does, from satellites to remote photography and internet communications, and nanotechnology, astrophysical instruments, cheap telephones, medical genetics, superconductivity, and nuclear fusion, to name a few. But we should not force technology to try to solve simple problems. It can always be adjunct to decisions, using common sense.

And, we don't really need that much more knowledge and technology. We are slow applying the knowledge we have now, and if it cannot promise a profit, it is applied even slower.

What should we do? Go back to the Neolithic? Live without scientific technology and suffer? No. But, before we race ahead, let us resurrect ways of applying knowledge to the base of life and not to the symptoms of a too-rapid, unconscious growth. I mean let us use 'common sense.' Common sense is usually defined as sound practical judgment that is independent of specialized knowledge and training—Or just plain ordinary, practical good sense. Common sense is easy to apply to our lives or civilization, and it has fewer unforeseen effects on health, ours and the planet's health.

What can common sense tell us to do about our health? Eat less, choose better foods, get more exercise, and so on. What about growing food or getting wood or other resources in nature? Don't use agricultural techniques (such as plowing) or chemicals that cause erosion or destroy the soil. Don't plant crops that rob the soil of all its nutrients. Don't harvest using damaging techniques.

How can common sense be applied to politics? It could set up circumstances for nations and people to represent themselves. Break up big nations, such as China, France, Germany, Brazil, and the US. Allow 'ethnic' and ideological small nations to join the UN. In each nation, limit officials to 5 years, with no donations and no reelections (including family members). Do not use violence to force your opinions on people. Guarantee basic income and health care, so people will not worry about it and can apply themselves to their passions. Pay unlikable jobs like butchery or garbage collection at higher rates than fun jobs like acting or playing sports.

How can common sense be applied to population size? By educating women and men, by licensing children, by rewarding limited reproduction. Don't use violence to keep people from planning their families.

How can common sense be applied to global problems? We could leave enough space (wilderness) for animals and plants to grow freely, even if it means setting aside 75 percent of the planet, as Paul Shepard recommended. Survey and monitor all populations. We could design our land, air and ocean transportation corridors to avoid interfering with the established migratory routes of birds, animals and fish. And, of course, we could clean up the ocean and atmosphere, removing plastics and junk from the first, and pollution and carbon from the second. We can see how to apply it, we just need to do it.

So, there it is. A simple outline for applying common sense. Cheaper, more participatory, with fewer negative effects, and resulting in healthier places in a healthier planet.

How could we activate this common sense approach? Should we limit television, to PBS or Education? Many things can distract us: Television, television in public places, television on computers and tablets. Not every person in every culture wants a television or the constant bombardment of violent or bad news. Should we offer everyone the chance to participate in thought experiments to improve the human, and nonhuman, condition? Yes, we should. The when and where are now and here. And, education can help guide our actions, with less drudgery and more fun.

Fun with Words Numbers Links & Images

Education is not just literacy, and that literacy, the ability to understand what words really mean, is not enough anyway to characterize education. Garrett Hardin points out that it needs to be supplemented by "numeracy," the ability to quantify information and interpret it intelligently—computers, remember, use numbers for everything—and, on another level, by "ecolacy," the ability to take into account the effects of complex interactions of systems over time, as things are interconnected and affect one another. Together, these are three major filters against folly that citizens can use against the blindness, short sightedness, and sheer idiocy that experts disguise as eloquence or expertise. With the increase in the complexity of civilization, it is necessary to recognize a fourth filter, "imagacy," the ability to work with complex patterns of visibility.

Literacy (Words). Literacy is the quality of being literate. Specifically, it is the ability to read a short passage and answer questions about what was read. Being literate is being characterized by learning, being cultured, or being educated. A person who is educated has been "led" from ignorance, out of the self in other words, by fostering the growth and expansion of knowledge through a course of formal study. But, most knowledge is concerned with survival first. It is important to know what plants to eat, where to find shelter, and how to make clothing. This was, and still is, the most basic level of literacy. Education should include a core of survival skills, and mathematics as a liberal education always has. Poetry and narratives should still be memorized, as well as being read on a computer.

Literacy, as the skill in the written and spoken language, enables readers to draw on the wisdom (and foolishness) of human beings distant in space and time. Hardin notes, in a discussion of the sins of the literate, that language has two functions beyond communication: "To promote thought and to prevent it." The second function is why literacy has to be accompanied by the ability to think critically.

Numeracy (Numbers). Numeracy involves the ability to measure and to interpret quantities, proportions and rates. Hardin warns that human beings have learned how to use literacy to hide numbers and the need for numerate analysis. He draws attention to the problems created by always thinking solely in terms of dichotomies, e.g., safe vs. unsafe or pure vs. impure, rather than in terms of relative risks and benefits. James Lovelock has also noted our inability to assess risks mathematically. The quantitative analysis that is so important in science, technology, business, and government is dismissed with indifference by many experts.

In a complex, rapidly changing society, understanding quantities, ratios, and rates and duration of time, is crucial. Numeracy has limitations, also—the conclusions of an accurate mathematical analysis are only as good as its premises. Intense interactions of people in larger concentrations also produce more information and data. Information in the entire world is estimated at over 1000 petabytes (or a quadrillion books of 170 pages).

Ecolacy (Links). Ecolacy was once achieved by studying natural history, the plants and animals that surrounded every human group. Ecolacy is the ability to follow the links or connections between things. This would allow the effects of the interactions of systems over time to be taken into account. Scientists have been extremely successful using reductionist methodology on every problem, breaking problems down into their components and studying the properties of these components and their interactions. This has led to the ascendancy of mechanical science—thinking that one can do just one thing. Hardin stated the important ecological understanding of ecolacy as: 'We can never do merely one thing." This statement is now known as Hardin's Law, and it means that there are always wanted and unwanted effects, products and wastes.

Reality is composed of causal chains of events rather than single effects. Events are embedded in causal networks and are produced by multiple causes and have multiple effects, each of which triggers a causal chain of future events. Hardin contends that since we cannot do just one thing we must always ask and answer that question: "And then what?" We have to try to ascertain the benefits and costs of proposed courses of action on both the individual, social, and ecological levels. The ecological systems way of thinking employs scientific theories and knowledge to study the interlocking processes characterized by many reciprocal cause/effect pathways. The systems way of thinking must become an integral part of thinking if we are to adequately control technology and human actions.

The world is too complex for our minds, suggested G. P. Marsh and many later thinkers. So, our minds have to filter out what is less important.

We filter data, arguments, emotions, and information. The filter allows a total picture of the whole with relatively little information. Of course that picture might be wrong. But it is clear, as a result of filtering and thinking. Other human activities act as filters, also, especially culture. Hardin contends that most of the major controversies of our time can be better understood as the result of the participants relying too much on any single one of these three filters. No one filter by itself is adequate for understanding reality and predicting the consequences of our actions. There is, however, a filter that has gone unmentioned, in its ability to concentrate or distort human perception. That filter is the image.

Imagacy (Images). Images can model the system in miniature. An image is an imitation or representation of something. It can also be a symbol or type, a metaphor or concept. An image can stand for something else, for instance the image of a dove is often used as a symbol for peace. In the etymological sense a symbol is something 'thrown together,' as a problem is something 'thrown forward.' Unlike an image, a symbol often represents some other thing, process or quality. Symbols can be processed by Analytical Engines or computers. These machines have been used for metaphors of the brain, which also processes symbols, that is, the operation of both parts of the metaphor, 'brain=computer,' can be described in terms of algorithms, or mathematical rules for manipulating symbols.

An image, especially a cosmology or an image of the world, models the system in miniature. From the image, which can be a paradigm or mind-set, a whole system can develop, with unique goals, rules, parameters, and structures. A cosmology includes a mythology constructed as a poetic system. Mythologies and religions can be understood as great poems. When recognized as such, they point through individual things and events to the ubiquity of a presence that is whole in each. This is what P. B. Shelley meant when he wrote that poets are the 'unacknowledged legislators of the world'—not that they pass laws or prophesize the future, but that they generate images for the future, and these images can be articulated into goals to influence our actions. Bad images, from indifferent poets, can relate to severe cultural problems, as when popular Italian poets romanticized the violence and hatred of Fascism. The mechanistic images of science, from Shelley to the Fascists, supported much of the violent conquest of nature.

Kenneth Bounding notes that the image as a cognitive construct of the world has several aspects: Spatial, temporal, personal, relational, value, and affection (emotional) for each individual. The total sum of individual images is a world of interrelated constructs. This parallels the experiences of other living beings. Using its senses, each participant creates an image of nature, or world—*umwelt*, life-world, is the term used by Jakob von Uexkull)— from the sensations that are meaningful to it, and which limit it. Simple beings, such as bacteria, make a relatively simple image, whereas more complex beings, such as apes and humans, forge more complex images. In fact, human beings design complex images for a variety of purposes, including to make a profit or to persuade others to have fun with words.

Asking Questions to Educate Ourselves

Questioning can lead to starting a thought experiment. Questioning, has a long provenance in human experience. The Greek philosopher Socrates was one of the more renowned questioners. He approached teaching through a disciplined, rigorous dialogue with people he met on the street, sometimes by accident or often by design. Socrates tried to get others to recognize the contradictions in their ideas; he assumed that incomplete or inaccurate ideas would be corrected during the process of questioning, and hence would lead to progressively greater truths. He never seemed to reach an end to questioning, however, perhaps because by itself questioning could only do so much with definitions and concepts. His method was a common search, through conversation, for the goal of truth.

Socrates asked questions as part of a conversation with others. He seemed concerned with discovering what the opinions of others were based on, an invisible truth that could be made visible with questioning. The questioning forced the other participants in the conversation to try to agree on the truths beneath the opinions. Socrates professed ignorance of the truth himself, in fact or in pretense, ignorance being the first step in the pursuit of knowledge. The process of questioning subjected opinions to real examples from real experiences, an empirical method leading to a more general concept. Through a thorough questioning Socrates demonstrated that knowledge was quite often uncertain. Questions were also meant to examine life as well as belief and truth, and to show that often people were ignorant of their ignorance. Socrates held that disciplined questioning enabled the other to examine ideas logically and to be able to determine the validity of those ideas.

The ecologist Garrett Hardin used questioning to illuminate partial knowledge and to track connections between things. Questions establish the limits of assumptions and perspectives. They can clarify the focus of a problem and test evidence related to any problem. Questioning can be used as a device to focus on a specific problem, not only the extent of the problem, but its aspects. Questioning can also be used to explore specific aspects of the dimensions of thought. Of course, questions can refine the process of critical thinking and can allow refocusing in a wider or narrower context. This type of questioning arrives at answers as workable hypotheses or guidelines for making decisions about operating in the world. Without certain knowledge, however, we can make adaptive decisions, based on partial knowledge. His questioning took on the form of asking what happened then, after an answer was arrived at: "And then what?"

Questioning works in a conversational way by weaving ideas together. Konrad Lorenz decided that humanity would indeed have destroyed itself by its first inventions, were it not for the very wonderful fact that inventions and responsibility are both the achievements of the same specifically human faculty of asking questions.

More than an annoying part of a conversation, questions are legitimate ways to approach a known or unknown situation. More than just a way to turn around a conversation, questions are tools that allow you to surround a topic and define it more completely. More than simply an admission of ignorance, questions can form a phenomenological spiral that allows you to return to a subject from different perspectives with different levels of understanding.

Questions widen a narrow field. Hardin points out that concerns about narrow issues, such as pollution, can cause a deep examination of the process, such as distribution theory, that cause or contribute to the issue. Human activity simply produces things that we want and things that we do not want, such as pollution. As we ask questions about who pays and who benefits, we are able to think or rethink about these things.

Questioning can get to the basis of any conversation. But, questioning can also frame and direct the conversation. Questioning also provides feedback for any answers. Therefore, questions will be a critical part of thinking. In terms of stimulating learning and creativity, questions are sometimes more powerful than answers. What do you think?

Conceiving Big Thought Experiments

Thought experiments modeling global dynamics, as well as modeling urban dynamics, are faced with significant challenges. Other thinkers and authors conclude that we cannot fully grasp the trajectories of urban systems, nor the planetary scale impacts of industrialization, nor the social ecological feedbacks, nor the ways in which the changes to the environment will then affect the urbanization process itself. Early models in the 1950s were concerned primarily with local areas or regions. After 2011, the first global models of land-use change emerged. A key feature to address is the integration of human and natural systems, which arrive from the fact that we are faced with feedback loops between cities in the global environment. These loops occur in a parallel and simultaneous fashion. While urban systems and processes have an effect on the local environments on a massive scale, changes in the global environment affect urban areas differently and at different rates. Very few models attempt to close the feedback loop by addressing the effects of changes in the systems on urban system.

Urban growth models make projections about future population growth and physical expansion of the GDP without paying attention to information about unintended costs, risks or uncertainties that arise from the actual project. But, unless we develop integrated models to address multiple scales of interactions or responses, nonlinear to directories, thresholds, the importance of economic agency, and the role of incentives and prices, our capacity to fully explain and realistically project how changes will affect urban growth and social form will be extremely limited.

We have a poor understanding of the needs and management of urban ecosystem services and large areas in South Asia, Africa and Latin America, which are critical areas facing rapid development and are greater threats to protected areas and hotspots. In Africa for instance the novel structures and human fluxes associated with urban are understudied.

Urbanization itself is a slow variable through which changing land cover, pollution and depositions may increase vulnerability to disturbances. At the same time urbanization may lead to higher frequency of disturbances through impacts on climate change. Urbanization therefore represents a complex interaction between slow and fast variables which need to be addressed in order to understand how their responses are linked.

We can design places as organic wholes to promote the well being of individuals and the common good. But, we can only really do so as participants of ecosystems (and this takes us back to the fundamental lessons of physics: That we cannot *not* be part of an experiment, and that disorder creates order which creates disorder). We need to recognize that they automatically participate in everything. Furthermore, due to the uncertainty in dealing with large, long-lived systems, we have to learn to accept that the system needs most of its own productivity and to limit our use of the systems below critical limits, within the flexibility of the system.

Community engagement with designers and designs can be encouraged through cognitive dissonance. And that can be created with notices, posters, events, and prizes. Participation encourages enthusiasm for the project. Performance design creates more empathy for other participants, where the affects of actions are anticipated and the emotions of others are experienced. Community and online networks allow observation of various kinds of consequences under different conditions.

All are cultural experiments. None have three-million-years of trial-and-error experience, and the little experience some 'Younger Dominator Cultures' do have indicates that they will always eventually self-destruct, usually within 200 years.

Urbanization was a new experiment that replaced the tribal and village experiment. And, urbanization has been complicated by the energetic exuberance of the industrial experiment, running coterminously.

All we have now is experiment, so we have to experiment to create new cultures that are not doomed to fail. Maybe we have enough pieces of Indians, Jews, Brazilians, Asians, and others, to try. And, enough knowledge, science and common sense to mix in.

Part 1: Animals & Pets

Experiment: *Buffalo Hides*

For a first experiment, let us trace the history of and a future for bison. Traditional forest Indian clothing was made from sagebrush, shredded cedar, and willow bark. Tanned deer hides were used for ceremonial clothes, although after trading with Plains Indians, styles of other tribes began to change towards buckskin clothing and moccasins; breechcloths, leggings, and shirts for men and dresses for women. Western and eastern Indians also traded for buffalo robes.

I have been so entranced with bison—not the water buffalo in southeast Asia—but with the history of massive herds covering the continent, with the wild myths, even with images on coins. Where are the buffalo now? Where are their hides, their bones?

Think about it. Probably a billion buffalo were killed in the 1800s. Over 75 million hides were shipped to the urban east to American dealers. Where are they now? In attics, as robes or hats? Lost? Degraded to dust? Will we be able to save the last photographs?

What about the beavers? Where are their hides? In hats? Who of you has a beaver hat? What about wolves, otters, passenger pigeons? Well over 5 billion passenger pigeons were killed. Are their feathers in hats? Pillows? In vases for display? Or has all that degraded, also?

Have we forgotten the tall grass prairie and the short grass prairies, with their bison and marmots? Most all of the large animals were killed. We have a few small experiments in Illinois and Kansas that show we could restore the entire piss-driven prairie if we were committed to restoration.

Even today, states sponsor wolf or bear hunts, so that private hunters can kill for trophies or for amusement. And the hunters track and feed wild animals for months before the "season" opens, so that killing with lighted trucks and high-powered, automatic guns will be made easier, although much less "sporting." Sometimes the limits are exceeded on opening day. Some bears and wolves live in preserves, but cannot avoid sanctioned kills. Biological diversity isn't just being eroded, it's still being killed!

Bison have been coming back in private herds. Can we make a Bison Commons, as has been proposed elsewhere (see Dave Foreman, et al.), that would run from the Canadian plains to the Gulf of Mexico? The bison would need to have at least 3.2 million acres to preserve a fully functioning ecosystem, a minimum viable area for the herds and for the other species that lived with them. The system would have to be large enough to survive fire and droughts. Then we might be able to exploit a percentage of them wisely. And, then we might have a few bison robes and beaver hats again. Maybe we will value them this time.

Experiment: *Do We Need Wolves For Our Own Health?*
Think about whether wolves are important in an ecosystem. The North
American whitetail deer population is unsustainable—an indicator of
ecosystem instability. Deer diseases, such as Chronic Wasting Disease (CWD),
are becoming increasingly communicated to livestock and other species,
including humans. CWD has been out of control in the United Kingdom.
It could rapidly become out of control in North America, Africa and Asia,
and kill hundreds of millions of livestock.

However, it could be stopped biologically through wolf predation.
Wolves play an integral role in maintaining the health of wildlife and
ecosystems, and indirectly, livestock and public health. Recognition of this
role and its ecological consequences calls for greater respect, protection and
increased numbers of wolves in appropriate habitats across North America.

Many ranchers, farmers and landowners have recognized the role of
predators and other wildlife species in helping preserve healthy ecosystems
around and within their lands. We need an ecological approach to wildlife
management, with careful predator control policies and practices. This
means that the wolf, long reviled by cattlemen and sheep ranchers, and seen
as a competitor to be exterminated by many hunters, may be the savior of
America's livestock industry and arbiter of healthier deer and other wildlife
numbers. This means more wolves (and also cougars, bears and lynx) are
needed in deer and elk habitat. Predators could also indirectly control other
diseases, such as Lyme disease, harbored by rodents and deer.

We need a revolution in state and federal wildlife and natural
resource management. The adoption of principles and practices that
enhance biodiversity and healthy ecosystems, domestic, commercial
and wild—with greater protection for wolves—is an integral aspect of a
more enlightened, science-based approach to a better environment for all
beings. This is the core principle of the One Health movement, promoted
tirelessly by Dr. Michael W. Fox, that is now being embraced by medical,
veterinary and global authorities and agencies. This movement is based on
the fact that human health is based to a large extent on animal health, and
animal health depends on a healthy environment, on healthy, self-renewing
ecosystems. And, the holistic perspective of conservation medicine must
include increased predation by indigenous carnivores, such as wolves. We
need the whole healthy system.

Figure 22. Balkan Wolf (Credit: Balkani Wildlife)

Experiment: *Inconceivably Massive Extinctions*

Are there too many extinctions happening? The earth can be a dangerous place to live. But, death and extinction are part of life; they help create diversity. That is, diversity is caused by speciation and extinction processes through natural selection. Healthy ecosystems reflect a balance of processes, including extinction and colonization. Colonization is necessary to equalize local extinctions. The loss of colonization, for any reason, can allow species in a local system to be depleted. Island ecosystems, surrounded by poor areas, may lose species without colonization. If the matrix is rich enough, there should be successful colonization. However, extinction can result from runaway feedback in a system.

In the history after the Cambrian explosion of life 550 million years ago, there have been five great catastrophes resulting in the extinction of over 75% of all species at each time. In the worst extinction event at the end of the Permean era 250 million years ago, volcanoes in Siberia pushed vast lava flows across land and into the ocean, blowing megatons of CO_2 into the atmosphere, changing the climate. Lava affected ocean temperatures, possibly releasing clathrates (frozen methane). Afterwards methanogenic bacteria blew out megatons of methane, another greenhouse gas. As temperatures spiked, the ocean stagnated and became more acidic, releasing hydrogen sulfide. Likely, 96% of all species suffered extinction, including the tabulate coral ecosystems. After each extinction, the system took about 5 million years to restore biodiversity to pre-event levels.

Even now in middle age, the planet and the activities of its species may threaten many other species, including the human, with death. This event, occurring now at your local wild areas, is the Sixth Great Extinction.

Big Problem: Massive Anthropogenic Extinctions. Although it may be as extensive and emotionally scary as the first five, it is not being initiated by global Ice Ages, planetary collisions, atmospheric poisoning, or variations in solar radiation. Instead, our decisions to be greedy or remain unconscious have scaled up into accelerated climate change from pollutions and from the conversion of complex ecosystems into impoverished ones.

Currently, 100 species per million per year are entering extinction. There is now a high degree of certainty that millions of species will reach extinction in the next 100 years. Temperatures of mountain habitats are easily measured, as are conditions that mountain species can tolerate. As climate warms, cold-tolerant species have nowhere to go but up, and we know the height of the mountains. We know the planet will heat by 2° F this century, or 5° F, if business and industry are conducted as usual. Chris Tomas et al. found that at the lowest degree of global warming (1.4-3 F) about 18% of species are 'committed' to extinction. At higher rates the number goes up: 3.2-3.6° F then 25%, and over 3.6° F then 33%.

Untimely extinction is a catastrophe. Normally, at the evolutionary time scale, extinction is balanced with speciation; but, human destruction of habitats has increased extinction thousands of times. Forcing species to become extinct is immoral and reprehensible.

Michael Soule identified 4 major agents of extinction: Overkill (the cause of most traditional extinctions); Interference, or habitat destruction and fragmentation (the future of extinctions); Introduction of Exotic species (especially those introduced to islands); and Chains of extinction (as a ripple effect through disrupted communities).

Unfortunately, our civilized activities are forcing the climate into a new warmer state. Our pollution is assisting the acidification of oceans. Our interference with fish populations, our destruction of habitats, our conversion of wild landscapes, and our dumping of novel chemicals and materials—are creating another perfect storm threatening much of life!

How many extinctions can the planet take at once? Will the extinction spasm of millions of insects and fungi have a greater effect on forests than other extinctions (such as Pleistocene megavertebrates)? Will a major (90-98%) loss of forests in developing countries be more risky than it was for Europe and North America? Once again, thousands of species are going extinct as their habitats are modified by large-scale human activities.

Big Solution: Massive Human Actions. On a large scale, we must stop land conversion. Preserve large areas for wild ecosystems. Simple size may not be enough to preserve a habitat if the original biodiversity is impaired. So, massive restorations will be necessary. The size should be large enough so that species will not be vulnerable to "extinction vortices" caused by genetic or environmental stochasticity. In this area, disturbance from resource reclamation or recreation would probably be the greatest threat.

But first, we have to make informed decisions based on better values and knowledge. Emergency plans have to be made quickly. Actions undertaken now will determine whether the planet will be biologically impoverished for the next 5 million years. There will still be dangers from plate tectonics, climate shifts or collisions with planetesimals. Life suffers mass extinctions, but it responds with larger equilibrium levels of diversity. For each instance of loss, we have to save everything that is restored.

We need to reduce the human footprint, reduce loss of biodiversity, and increase it with cautious restoration. And, of course, we need common sense about conservation: Use less, consume less, and produce more on less. We have to anticipate losses and their consequences. If people ignore the need for emergency actions to slow climate change, from conservation to new technologies for alternate energies, and new designs for sustainable patterns, and for wildlife conservation, then threatened and endangered species will soon become extinct. Many other linked or dependent species will go. If this happens, human beings will lose much of great value that cannot be replaced. A planet of lonely humans and vermin might not last long.

Experiment: *Are Wild Dogs the Ancestor of Wolves?*

Wolves originated in North America and moved out to Europe and Asia. Humans originated in Africa and moved through Asia and North America. Somewhere, within 3-4 canine generations, dogs developed separately from wolves in eastern Asia; they both came from a common dog ancestor, according to Janice Koler-Matznick (*Dawn of the Dog*, 2016), and then spread through Europe and North America. Many of the dog species associated with human communities as scavengers, but some stayed wild, and are still wild. Scientists have speculated that the original dog may have resembled dingos.

Like domestic dogs, wild (or free-ranging) dogs manifest themselves in a variety of shapes, sizes, colors, and even breeds. The primary feature that distinguishes wild from domestic dogs is the degree of reliance or dependence on humans, and in some respect, their behavior toward people. Wild dogs survive and reproduce independently of human intervention or assistance. While it is true that some wild dogs use human garbage for food, others acquire their primary subsistence by hunting and scavenging like other wild canids, such as coyotes, jackals, and African wild dogs.

Wild dogs are usually secretive and wary of people. Thus, they are active during dawn, dusk, and at night much like other wild canids. They often travel in packs or groups and may have rendezvous sites like wolves. Travel routes to and from the gathering or den sites may be well defined. Wild dogs are highly adaptable, social carnivores. Most are about the size of a coyote or slightly larger.

Many breeds of dogs are capable of existing in the wild, but after a few generations of uncontrolled breeding, a generalized mongrel tends to develop. Often, it has a German shepherd or husky-like appearance.

Feral dogs, that is, dogs that have been abandoned or escaped to the wild, are sometimes a problem to wild and domestic dogs. The survival of feral dogs, much like that of other wild canids, depends on their ability to secure food. These feral dogs are sometimes adept predators. In areas where they have not been hunted and trapped, feral dogs may not have developed fear of humans, and in those instances such dogs may attack people, especially children.

Clearly, the impact of feral dogs, both on livestock and wildlife, varies by location and is influenced by factors such as availability of other food, the number of dogs, and competition by other predators.

To prevent damage from wild or feral dogs, many of the same strategies used for wolves or coyotes can be employed: The use of guard dogs (or donkeys); fencing in domestic livestock; mounting loud auditory devices, sound-activated, from fenceposts; spraying chemical repellents around the boundaries of livestock areas; and, also trapping repeat attackers with steel leghold traps.

Another important strategy is the proper care and confinement of domestic dogs, which should be licensed to the owners. Dog breeding must be controlled. Unwanted dogs should be placed for adoption or destroyed rather than abandoned. Public education is an important first step, with laws to handle destruction and disposal of unwanted animals.

Aboriginal village dogs are ancient indigenous populations of dogs, also called "landraces" for their adaptation to the local environments (Indian pariah dog, dhole, Australian dingos, and arctic breeds, among others), whose movements and reproduction have been under no or very little direct human control. These landrace species need to be saved from extinction, not by breeding, but by restrictions on breeding, by decreased artificial selection and by preservation of their wild territories. Some free-breeding aboriginal dogs are still pure, not mixed with recent dogs from Europe and Asia. But, with the exponential expansion of human populations, and their economies and domestic animals, these last surviving original dogs may be lost.

Experiment: *Are Sick Cats Are Destroying Wildlife?*

Cats have earned a reputation as relentless killers of wildlife. They have been named among to the top 100 of worst invasive species known. There are 47 million pet cats in the United States and an estimated 50 million more feral cats (HSUS data). The total world population may be 600 million.

Cats can range from 2 kills a week to 30 kills a day; only a fourth are brought back home. In Australia 5-20 million feral cats catch 5-20 billion native species per year. Cats are credited for countless island extinctions. Cats can claim 14% of modern bird, amphibian and mammal island extinctions. Felines accompanying their human companions have gone on to prey on the local wildlife, and they have been blamed for the global extinction of 33 species, mostly on islands (U. Nebraska), including New Zealand's Stephens Island Wren. On land, cats are guilty of more deaths and extinctions than was thought previously. Feral cats and strays—not house cats—are responsible for the majority of the killings.

A study by the Smithsonian concludes that between 1.4 to 3.7 billion birds lose their lives to cats each year in the US. Around 33 percent of the birds killed are nonnative species. And, between 6.9 to 20.7 billion small mammals succumb to the predators. In urban areas, most of the mammals were rats and mice, though rabbit, squirrel, shrew and vole carcasses turned up in rural and suburban locations. Just under 70 percent of those deaths are by unowned cats, about three times the amount by domesticate cats.

Cats may also be impacting reptile and amphibian populations. Based upon data taken from Europe, Australia and New Zealand and extrapolated to fit the United States, some scientists suspect that between

258 to 822 million reptiles and 95 to 299 million amphibians may die by cat each year nationwide. These estimates, especially for birds, far exceed any previous figures for cat killings, and also exceed all other direct sources of anthropogenic bird deaths, such as cars, buildings, towers and wind generators.

This is a catastrophe for wildlife, an emergency! And the source of the emergency is human beings, pet breeders, pet owners, and pet protectors (feral cats are most often the offspring of domestic pets). Although owned cats may have relatively less impact than unowned cats, due to their 'inside' time, owned cats still cause substantial wildlife mortality.

Programs in which feral cats are caught, "fixed," and released back into the wild unharmed are undertaken throughout North America. Often without thought for native animals and without widespread public knowledge. While cat lovers claim that these methods reduce wildlife mortality by humanely limiting the growth of feral colonies, this is a weak assumption. The community dumping of cats amounts to abandoning highly stressed and immunocompromised cats to risk; this is an act of cruelty and irresponsibility, from animal welfare, wildlife protection and public health perspectives. Cat colonies should be a wildlife management priority. Many colonies may need to be exterminated to protect wildlife.

Local cat numbers can quadruple within months after a mouse or rat outbreak. If the prey population crashes, the cat population follows. Cats starve and succumb to diseases like toxoplasmosis. The cats kill all the birds, then, the cats all die from toxoplasmosis. Cats, however, are the only animal in which the parasite can complete its life cycle. Cats become infected by *Toxoplasma gondii* by eating the immature forms of the parasite contained within the muscle or organ tissue of other infected animals, such as mice. Those immature forms, or cysts, mature inside the cat's intestines and are excreted in the cat's feces. If another animal—including a human—inadvertently consumes those feces, they become infected with toxoplasmosis. The domestic cats do not fit in most wild ecosystems. They tend to degrade the systems they dominate.

There are actions that can be taken. Ban the feeding of feral cats. Kill as many feral cats as possible. Keep all pet cats inside, or in an enclosure, or on leashes outside. Reduce outside time. Reduce the number of pets. Overpopulation may need more drastic, temporary measures. Euthanasia alone will not solve problem of all feral cats.

Overpopulation is a 'man-made' problem. And, it is an ethical nightmare. The interests of all stake-holders must be considered as well as the consequences of actions. We must take decisive, possibly repugnant, actions to correct this awful imbalance and to avoid a larger tragedy.

Experiment: Will Running Wild Horses Survive?

In the last forty years, fifty percent of the world's wild species have been decimated, extinguished. More wild horses and burros live in short-term holding pens than exist freely in the wild. These captured horses are being supported by taxes rather than grasses. There are still over 55,000 wild horses and burros on the range in 2010. A recent amendment to the Wild Free-Roaming Horses and Burros Act of 1971 (WFRH&B Act) by Senator Burns, who slipped an omnibus clause into it without public input, knowledge or approval, states that captured wild horses and burros can be sold 'without restriction.' Those last two words changed the intention and purpose of the WFRH&B Act and opened the door for "kill buyers" to purchase the once free roaming wild horses and burros for slaughter and eventual human consumption.

The Bureau of Land Management (BLM) bases the removal of the wild horses and burros on what they have named Herd Management Areas (HMA) and Appropriate Management Levels (AML). The BLM and the Department of the Interior (DOI) determine these numbers by some mysterious computation unknown to scientists. The government continues to ignore and violate existing laws, in place for years, to protect the wild horses and burros to live freely without harassment. The wild horse advocates do not have a lobbying group or much financial strength. Most donate from their own bank accounts to support needed legislative changes. It has not been enough.

Decimation of wild animals is still going on now, taking its toll on the wild horses and burros roaming freely on the ranges in 10 US states, by the BLM under the DOI. These are the government departments that are suppose to manage the public lands for diversity and multiple use. These are the departments tasked with protecting and preserving the wild horses and burros and their freedom to exist on public lands, unharassed, according to legislation passed in 1959 (The Wild Horse Annie Act of 1959, TWHA Act) and 1971 (WFRH&B Act). Let's enforce those Acts, so that wild horses and burros, acclimated to the environment after 500 years, can continue as part of wild ecosystems.

Figure 28. Toma, a rescued horse.

Experiment: *Why Do We Need Domestic Animals?*

Domestic animals have accompanied human groups since the end of the latest ice age. Domestication was simply the latest stage of following, trying to control and killing wild animals for food. This change occurred in the same era as the domestication of wild grasses. It allowed a more constant source of high-quality protein.

Are all domestic animals to be treated the same? Let's look. There are food animals, such as beef cattle, sheep, chickens, dogs, and pigs. Other parts of the animals are used for leather or wool. Food animals tend to have high-end diets for fattening for market.

There are working animals, such as pigs, horses, dogs, and birds. These animals are used for hunting, transportation and finding mushrooms, using senses that are more expansive than human ones.

And, there are companion animals, such as cats, dogs, birds, and horses. For the most part these animals live in or near human habitations, entertaining or helping their 'masters.' They are fed extremely well, sometimes sharing the serving table. They can also provide psychological services for their owners, many of whom live alone or are socially challenged by human interactions.

What would be the alternatives to domestic animals? The first answer could be machines, which have already replaced dogs, mules, horse, water buffalo, oxen, and others as beasts of burden. Robots are providing medical care to people in some hospitals.

A second alternative, relating to food animals like cattle, would be to return to hunting wild animals, such as deer, bison, antelopes, and others. Although the scale would be really problematic. Reducing the vast numbers of feed animals on feed lots would happen only if diets changed to be more vegetarian. Game or Bushmeat has been successful at small scales, and in fact, has worked with international desires for wild animal meat; Flying Fox bats for example are canned in Australia and shipped to Australians everywhere.

A third alternative would be to eliminate pets, such as horses, pigs, cats, dogs, goldfish, and the other exotic ones. This would problematic due to high human emotional attachments, from identifying with pets and needing pets for social support. Some people have replaced pets with virtual computer pets, which have proven to be as messy and frustrating as real animals. The majority of pet owners prefer the living animals, although steep veterinary medical bills may discourage some.

Although it may not be possible, or even desirable, to replace all domestic animals, these possibilities, including dietary changes and greater emotional health, may reduce the pressure on the millions of extra animals and on the environments that they pollute involuntarily. We should try to reduce the numbers of domestic animals, for reasons of health alone.

Experiment: *Can the Benefits of Pets Justify Having Them?*

Pets are a blessing and a problem. They require high maintenance, especially single dogs raised in confined urban spaces. They prefer rich diets, unlike anything a wild animal experiences. They are raised in a predominantly human environment, empty of normal stimuli. They are placed under human expectations that go against their wild behavior. For example, many animals do not like to be rough-housed as adults and have to inhibit biting behavior towards their owner.

Pet animals, especially horses and dogs, should not be raced. Although they like to run, sport running is hard on the animals—sled races with Clydesdales pulling tons for short distances at speed is a case in point. Greyhounds are often raced until their bones break (fortunately there is a society to adopt used Greyhounds).

If we are going to have pets, we need to provide the proper environment; for dogs, this means a pack with a good-sized area, and for horses a herd (minimize races and rodeos). Horses in grasslands would contribute to ecosystem health. Dogs in packs, however, do not contribute much to ecosystem health; they hunt in packs, but are relatively uncoordinated and less effective hunters; they tend to overstress animals they cannot catch.

What about cats? Cats are popular, despite being more wild than dogs or horses. They are also small, which makes them ideal for house animals. We should stop breeding them, especially in kitty mills. Perhaps we should end their owned plaything status.

If we are going to continue to have pets, we need to educate, test and license their owners. We would have to limit their numbers to so many per environment (or county or state); people may have to enter a lottery for the opportunity. Veterinarians would also have to be licensed and held to limited income for their services.

What would happen if we stopped breeding animals for pets. For decades, there would be animals suitable for adoption. After these died, humans would have to turn to each other for companionship, as it was for most of the history of the species. That would bring improvement to the lives of many.

Figure 30. Puppy Mill (Credit: Dr. M. W. Fox).

Experiment: Are There Benefits of Pests?

The pest killing industries present an image of a pest-free world, where toxic chemicals and poisons attack any unfriendly insect that dares to show its mandibles. Should we kill them all if we could? Or should we find a way to live with them, with just minor changes in our behavior?

Screens work well to keep out insects. Other slight changes, such as adequate ventilation, can control termites and fungi. Certainly, we could carry out the occasional bee, spider or bat.

However, recent press about disastrous blowback when humans target species deemed a nuisance, should give pause to the impulse to wipe out even the most bothersome of pests. Two examples. First, the 90 percent decline in the population of the monarch butterfly in the last two decades from spraying herbicide on genetically modified corn and soy in the Midwest, inadvertently destroying the milkweed on which the monarch caterpillar must feed. And second, the emergence of antibiotic-resistant bacteria from rampant misuse of antibiotics, both to treat viruses in humans and to fatten up livestock that are not sick. Consequently, people are at risk of picking up antibiotic-resistant superbugs by eating meat or staying at a hospital.

Houseflies are undeniably a nuisance as they carry many pathogens dangerous to humans, like typhoid, dysentery and tuberculosis. They consume whatever is available, from sugar to feces. Pathogens can be spread from contacting their body parts, vomit or feces. Houseflies evolved some 65 million years ago and live virtually everywhere but the Antarctic. Females lay thousands of eggs which hatch into larvae (maggots) that feed voraciously for a week on dead and rotting matter—like feces, garbage or carrion—before forming a pupae from which the adult fly emerges.

Thus both larval and adult stages provide the vital service of cleaning up all manner of decaying organic materials and returning the nutrients to nature. Both also figure heavily in the diets of many species, such as lizards, birds and spiders. Given these easily appreciated contributions to ecosystems worldwide, humans have adopted a largely live and let live attitude toward houseflies, though few would advocate doing away with their flyswatter.

In fact, engineers are looking to housefly larvae, with their lust for gorging on feces, to solve the problem of managing the huge volume of manure generated on pig farms. Pig waste is often spread on nearby cropland or forests after storage first in big lagoons, with the potential for polluting both soil and water from the pathogens and excess nitrogen in manure. Scientists propose to rear maggots in pig waste as an ecologically safer way to biodegrade it. Furthermore, before the maggots transform into flies, they could be harvested as a protein source for animal feed.

Mosquitos are considered the most dangerous insect to humans because they are vectors for transmitting infections like malaria, yellow fever, dengue fever, and West Nile virus, from person to person and are credited with causing more than a million human deaths annually. Furthermore, with mosquitos the attack can feel personal.

Mosquitos have been around for at least 45 million years and, of the thousands of species, only some are bloodsucking. Bloodsucking types variably feed on mammals, reptiles, amphibians, birds and even fish, and thus are vectors for animal diseases too, like heartworm in dogs and encephalitis in horses. Not all transmit diseases, however, and only the female sucks blood. The males feed on nectar, so they participate in pollinating plants.

The larvae of mosquitos are an important food source for fish species including bass, bluegills, catfish, guppies, piranhas, salmon, tilapia, and trout. Adult mosquitos are preyed upon by many insect-eating creatures, including dragonflies, birds, frogs, lizards, bats, spiders, and even other mosquitos. Innumerable other species have co-coevolved with and become dependent on mosquitos. Some biologists foolishly argue with an anthropocentric flush that mosquito eradication to extinction is a good idea: Disruption would be minimal for other species, and the world would thrive without disease-transmitting species; other benign insects would move into vacated niches. They think. Probably, the loss of a disease-vector mosquito species would allow another worse disease-carrying species to fill the niche.

Experiment: *Are Guard Dogs Better Than Payments?*

Unprotected food animals are often taken by wild predators. In many cases, the food animals are 'protected' by poisoned bait. The methods employed by state and federal agents to kill these predators, which include poison bait and 1080 cyanide guns, are often indiscriminate, killing non-target animals including endangered species such as the lynx and golden eagle.

This sanctioned adversarial attitude toward wildlife in general and predators in particular, then indifference about their suffering and disregard for their ecological value, adds up to unethical and non-sustainable ranching and farming practices.

The Bulgarian Karakachan dog has proven to be effective in many instances, protecting sheep, goats and calves from wolves, dogs and dishonest humans. Why should compensation be provided if guard-animals such as Karakachan dogs are not being effectively deployed? This would also save tax dollars in not having government agents go out on killing rampages to revenge losses.

The use of well cared-for guard dogs to prevent predation, as an alternative to lethal methods of predator control, is surely an ethical imperative for a nation of meat eaters. Government has been overzealous in killing so many predators that the land is filled with hungry prey that unbalance entire ecosystems. It is as sane and sensible as using goats, instead of Roundup and other herbicides, to control weeds.

Traditionally, guard dogs and shepherds form strong mutual attachments. This is good for both, and it is good for the profession of shepherding, which has been neglected in many countries. With so many animals in rural or wild areas, often turned loose for the winters or summers, shepherds are necessary, and more reliable than fences or map lines, and more economical in the long run than payments for real, preventable or imaginary losses.

Figure 33. Shara & Arul Bulgarian Karakachan Shepherd Training Dogs

Experiment: *Does a Porcupine Dream? Does a Pine tree Think?*
Yes, obviously. If you spend a lot of time watching animals and plants in a forest, you can see them think and likely dream.

Wthin their world images animals and plants, such as trees, consider options, then choose pathways–that is one definition of thinking. So, they think. Sometimes while they sleep or rest, images of actions form in the brain (or chemical memory); that is one definition of dreaming. So, they dream as well. Of course thinking and dreaming varies with the complexity of the organism. So, pines can think and porcupines can dream! They can compute, also. They process 'amounts' of things. Living beings have talents that we used to think only we humans had.

Adolf Portmann shows that every form of life appears as a gestalt, developing in a specific place. All living forms create an image of their environment. Genetics provides the proper image choices for some—frogs, for instance, focus most closely on objects that have the same size and trajectory as flies. Others must learn what is valuable.

Jakob von Uexkull suggests that the unfamiliar world of animals can be represented with bubbles to denote the self-world or phenomenal world of an animal. The world—life-image—is what has meaning for an organism. It is a focus. The first principle of the life-image theory is that all animals from the simple to the complex are "fitted to their unique worlds with equal completeness." Thus, we can understand a lot about animals by knowing their environment, and by how they interact in it, as well as what other plants and animals share it.

Experiment: *Can We Think Like Plants Or Animals?*

No. We cannot. But, we can approach other minds. We can understand their environment and their world-image as much as possible. The human place in nature can be explored through the concepts of anthropomorphism, anthropocentrism, and anthropometrism.

Human images of worlds were wedded to shape (anthropomorphism) rather than to position, measure, or language. Archaic peoples saw human forms in every form of nature. The order of nature was a human order. Natural events, like lightning or rain, had needs and reasons. And these events could be controlled by human rituals, which satisfied the needs or influenced the reasons.

Anthropomorphism is a necessary human way of knowing; all knowing is based on it. But the knowledge is not limited to just human experience. Anthropomorphism gave human beings a place in nature. Other beings were seen as relatives, acting and thinking like us, yet being 'others.' We understand other beings by expanding ourselves, not by shrinking them. This is one meaning of 'thinking like a wolf.' It is not 'wolfing like a wolf,' as someone once argued. It is remaining human while acting as if one had the identity of a wolf. This is how humans project experience into other beings, not make other beings into humans. The social lives of humans and other beings are not separate. Symbolic associations and transformations are made between entities.

Yet paradoxically anthropomorphism is limited. It is nonabstract. Humans project human need and thought patterns as guiding forces in nature. Nature, however, is not a human creation. Anthropomorphism is a personal interpretation of an order that encompasses all human orders. Anthropomorphism is limited by the range of human experience.

An increase in knowledge from the Neolithic to classical periods brought about a realization of the vastness and strangeness of nature. Concentration on human interests, anthropocentrism, resulted in greater success for the species. Humans became successful competitors. Then they became instruments of change in ecosystems.

Humanity is not the center of nature, however. It is not the reason for the existence of all. Humanity is one center among many, and that center is placed in context. The earth is not warped for human benefit.

The universe is not in the shape of humanity. Nor is it centered around humanity. But, it is measured and valued by humanity, as, indeed, it is measured and valued by all beings. Measure implies limit. The word 'morality' means to measure one's way. Measuring social interactions leads to moral codes and ethics. An ecological ethics can address the limited relationships of all beings without becoming entangled in the fuddle of reciprocity or sentience.

All three concepts dealing with shape, center, and measure are needed for human knowledge. By rejecting anthropomorphism, the experience of others is restricted, and the scope of self-knowledge is reduced. Narrowness of experience is a source of human insecurity. By rejecting humanity as its own center, the experience of selfness is suppressed. The self is the basis for exploration and success. By rejecting measure, perspective is lost. And if humans claim all value for themselves, the term is meaningless. The proper study of humanity is all beings, from animals and plants to bacteria and archaea.

Experiment: *What if We Formed Totems for All Living Beings?*
Enlightened modern hunters often form associations, such as Ducks Unlimited, to protect animals and their habitats from the accidental threats of modern civilization. A good hunter learns to think like her prey and to recognize its signs and requirements. Hunting traditions have taboos against taking too many prey or vulnerable individuals.

In the past, hunters revered their prey. Reverence for life is an ancient attitude, recently revived as a principle by Albert Schweitzer. True reverence for life makes no distinction between higher and lower forms. If we were to act so, distinguishing between pests and pets, we must do so in the sorrow of the recognition that we are killing by choice. True reverence for life entails reverence for death, since life and death are inseparable. Life should be used and shaped with respect, and experienced with awe, for underneath it is still an unfathomable mystery.

People have identified with animals for thousands of years, tracing their own ancestral lineages from them. In totemic religions, typical of foraging groups, each group is symbolized by a particular plant, animal, object, or phenomenon ('totem' is from the Algonquin word meaning relative). Clans are descent groups from these chosen or fictional ancestors. The first totem often came in a vision; it was used to coordinate out-marriages between groups. A totem, as defined by J. G. Frazer, identifies an intimate, special relation between humans and every member of the class.

For instance, the Arunta tribe in Australia is divided into totemic bands. Aboriginal band moieties in Australia divided people into two groups within an estate group: Half of society might be black, kangaroo, acacia tree, and goanna lizard, and the other half, white, emu, gum tree, and rainbow serpent. Such divisions made it simpler for marriage and unions. The rules of a totem also constituted good conservation practices: No hunting in sacred sites; prohibitions against hurting totemic animals; and, compassion to all animals. Totem animals (or objects or phenomena) could be contacted in Dreamtime (everywhen) when the invisible side of life was made visible.

In another instance, the Kwakwaka'wakw people in the Pacific Northwest of North America shared their ancestry with each other on a crest pole (sometimes called totem poles). They believed that each person had a special totem or guardian spirit, who bestowed some of her powers. The totem could be a plant, animal, substance, or event, such as lightning or cloud. These poles were commissioned by the head of the family to display status and ancestry. Ancestors included wolves, bears, whales, seals, sea slugs, and barnacles, as well as mythical beings such as Thunderbird.

The poles were not worshiped, but embodied beliefs about social realities, including descent, inheritance, power, privilege, and social worth. The poles were meaningful because they were the visible expression of the history of the family, which had to be publically recounted and witnessed. The construction of the poles was accompanied by the appropriate rituals, since each pole was owned by a supernatural being. The artist presented the faces only, so the stories had to be retold by the owners. The poles were also used as teaching devices, to help train children and spouses.

Totemism is a good way to have animals be represented in human society and law. It is more selective than pantheism or animism—the pantheist interprets the universe as holistic, but does not necessarily recognize any important duty to it; it is too neutral toward good and evil, destruction and preservation. An animist may not feel responsible if there is no distinction between life and death, if everything is merely transformed. But, the totemist provides a commitment to preserve the subject through the bond of brotherhood. Totemism teaches an invisible 'what' about the self, since it is an analogy between a social system and the natural world, in a religious context. People are parts of society, while people, animals and plants are parts of an ecological system. Empathy is accentuated.

Paul Shepard proposed an association for each species, although he mourned that machines had become the new totems, displacing animals as models of behavior. Independently, Michael W. Fox proposed a human association devoted to each species of wild being, so that every creature on earth would have a human constituency. Groups of people would form voluntary leagues dedicated to single species and their habitats, which are crucial parts. A feeling for participation in ecosystems is necessary. Leagues would have the ability to foster concern for nature, with a tendency to preserve it. A totemic disposition would increase a primitive awareness of the earth. There may be secondary or tertiary totems that overlap with other groups or leagues. Social networks and economic systems could be extended into the ultrahuman world by imagining that life and obligations were shared. Kinship could be reflected in totems.

Figure 36. Kwakwaka'wakw Crest Pole

Part 2: Forests & Water

Experiment: *Are We Removing Too Much Biomass?*

Ocean covers almost three quarters of the planet, and we live in an ocean planet. The ocean invaded land as algae, plants and trees. Trees are literally standing water. Over three fourths of the active biomass on land is situated in trees in forest ecosystems.

The structure of ecosystems is based on constant material and energetic exchanges. The matter present is called biomass; the material output is primary productivity. More mature systems have a richer structure and a lower productivity per unit biomass. There are more steps in the trophic pyramid. There is higher efficiency in every relation. The loss of energy is less, so less energy is needed to maintain the system. This is why old growth forestsfor instance, are efficient.

The net productivity equals the sum of stems, leaves, flowers, fruits, roots, and insect loss, minus the initial seed sown. The total amount of organic matter, or standing crop, is the biomass. Biomass accumulation is the increase in community organic matter as the difference between the gross primary productivity and the total community respiration. This difference is also the net ecosystem production, which is usually less than the net primary production. In mature communities, the net primary production may be high, but the net community production approaches zero; the gross primary production and total community respiration are almost equal. A good example of a mature community is old-growth forest.

For this reason, human intervention into climax or mature systems is usually detrimental. In early successional stages, there may be larger net community production totals. These can be harvested by humans without much damage to the system. One reason that gross productivity cannot be harvested by humans is that it would destroy the ecosystem; the wealth of the system would not be sustainable.

Almost every food web is ultimately dependent on the amount of plant tissue or biomass available for consumption. Biomass seems to increase with diversity. There is also an increase in the proportion of inert matter, and an increase in structures like paths and burrows. In an exploited system, diversity drops and the ratio of primary production to biomass increases. Mature systems can regress to earlier forms when exploited; but, then new species can form.

Forests have high rates of production compared to most other ecosystems. Forests have the most biomass, but the soils are not as deep as grasslands or swamps. As the forest ages, it undergoes changes in productivity, structure and population selection. Almost one third of the production and the biomass is underground in roots. Often root biomass is not included, although of course it should be. There is no satisfactory way of assessing root biomass, other than assigning it a percentage of above-ground growth. There are other problems accurately assessing biomass: Grazing by herbivores may have occurred; grazing by insects is almost impossible to measure; the trees may have shed leaves or limbs.

Forests change because of disturbances. Disturbances may be external or internal (although this distinction is artificial because of the interdependence of forests and climate, soils, streams, and animals). The variation of climate, for instance (an external disturbance), can change the physical and chemical parameters of the forest. Acid rain causes biochemical and soil chemical reactions, which in turn provide information to other organisms, which use it to regulate their physiological processes which can effect the forest, e.g., insect damage changes the level of seasonal growth of trees. Damage to trees from pathogens (disease agents) is an internal disturbance. Fire and biomass removal (mostly by humans, but also by streams, elk, etc.) are external disturbances.

Exploitation means 'use' basically, while harvesting refers to biomass removal. It has been argued that clearcutting (the removal of all or most trees and biomass) mimics a crown fire, or a hurricane, or an epidemic, but none of these activities strike all of the trees in a forest or remove their biomass from that location. The dead trees are still nutrients and continue to contribute to the system.

How much wood can be taken from a forest? How much biomass can be taken before the system collapses? Many terms in ecology, such as species, biomass, stability and diversity, are inexact. That means they can be measured only at one point in space and time, and the numbers are always uncertain. By taking too much of the biomass, within too little renewal time, the forest and other dependent ecosystems are doomed to collapse.

Figure 38.
Altazor Forest &
Noname Stream

Experiment: Promoting the Health of Forests

Historically, we have used forests without regard to their continuity or to their health. Partial knowledge and sophisticated technology has allowed us to exploit our environment beyond what is desirable for us or for other species. Moderate exploitation is necessary to live, but too much is unwise. A wise use of resources would not make the world less habitable. We are part of the system and must protect its health as a whole.

What is health? The absence of disease, doctors used to say, the traditional *via negativa*; it is not having cancer, infections, high blood pressure, diabetes, or other ailments. The World Health Organization (WHO) defines health as a total physical, psychological, spiritual well-being of a person.

The opposite of health is disease. But, what is a disease? Many diseases, at some stage or concentration, obviously compromise health. But some human diseases, such as sickle cell anemia, make one healthier in the sense of being able to resist another worse disease, G-6-PD deficiency (glucose-6-phosphate dehydrogenase), a genetic disease that causes red blood cells to dissolve.

In his theory of the implicate universe, David Bohm proposes that health is a result of a harmonious interaction of all the analyzable parts that comprise the extricate order—cells, tissues, organs, the body—with the surrounding larger environment. Health is a quality that is grounded in the total order of the environment (or implicate order). Health is a dynamic quality of the entire movement of the environment (holoverse) as it flows. As organisms sometimes interfere with others or with the flow of change, the harmony breaks down—we call that disease. Health is the dance of bodies that interpenetrate (to use Paul Shepard's image).

None of the bodies are completely independent or completely bounded; they are interdependent and open systems. A body is only maintained by a flow of energy and materials from its environment—much of this flow. Each of the animals involved performs a function in the forest. Porcupines often thin crowded nurseries of young trees. Mice also feed on the pupae of the larch sawfly, limiting those insects; mice also contribute (involuntarily, of course) to the health and balance of predator populations, from wolves and coyotes to snakes, hawks, and owls. Birds may eat a lot of seeds and make a lot of holes in trees, but they also disseminate seeds and eat insects and rodents. Insects may invade forests; their activities may kill trees, introduce fungi, and increase flammable material. Insects are also beneficial to plants, contributing to pollination and soil fertility. For example, aphids remove surplus sugar from plants, which when it gets to the soil adds to nitrogen fixation and promotes soil fertility; removing sugar also increases plant fitness in some kinds of plants.

Now we can address health in forests, which are larger entities made up of other beings—larger patterns made up of smaller ones.

A good metaphor for forest health is human health, which has been studied for thousands of years. Our first clue for unhealthy people is abnormal behavior—not moving or breathing, for instance. The first thing we do is classify the symptoms. Then we measure vital signs: Heart, blood pressure, and temperature, which can indicate a treatment. Health was also defined generally as the absence of disease.

The problem with the health of forests is that we do not have the same huge compendium of diseases and symptoms. In fact, we are just starting to compile the stresses and causes of forest illness. Furthermore, we tend to use our human values to judge health in forests; for example, we tend to think that forests should exhibit regularity, but many forests, such as boreal forests, are arrhythmic, that is they are punctuated by surprise events (as C. S. Holling suggests).

Human medicine itself is changing from being disease-driven to wellness oriented, from focusing on symptoms to describing the properties of health, such as interconnectedness and self-realization.

Health is a dynamic measure of ecosystem organization, vigor, and resilience. Organization is described by diversity and connectivity; vigor is related to the amount and speed of productivity; and resilience is a measure of reaction to stress. Too much stress, for example, leads to unsustainable patterns of behavior; continuous stress leads to a breakdown of processes that becomes irreversible—the system dies. Trees in a forest are always dying, either individually or in groups. Forests also may die, if enough of the trees die as a result of catastrophic change or too rapid or constant a change.

As a forest ages, the probability of insect and disease outbreaks increases. Yet, even catastrophic disturbances like hurricanes rarely damage more than 5 percent of a forest. More than being agents of mortality, insects, diseases, and animals are native components of complex food webs in ecosystems that contribute to the selection of certain kinds (including healthy) and ages of trees (determining the composition of the forest, which changes over time). Mammals and birds disseminate seeds. Insects pollinate some trees and overwhelm others (rarely more than 1 percent of a forest). Diseases remove stressed trees (also probably a low percentage on the order of 1 percent). Their effect on the long-term health of a forest can only be regarded as positive.

It is quite likely that harvesting a limited number of trees in a pre-dominantly natural forest could reduce the probability of disturbance as well as stimulate new growth, although it may equally possible that the regular removal of trees for human use may result in the gradual and long-term decline of a forest. This should be part of a very long-term study.

Experiment: *And Then What? Forest Equity & Balance*

Foresters are surely right to say that ecologists are premature doomsayers or Cassandras (for spurning the love of Apollo, Cassandra was given the gift of being always right, but never being listened to). The ecologists are certainly right to say there will be shortages, even if the dates and extents of the shortages are unknown—the function of a modern Cassandra is to be wrong. The warning should change the behavior.

Garrett Hardin used to say that the essence of ecology was found in the question, "and then what?" meaning that everything you did had a primary effect (there were no side-effects) on the system, every action a re-action, or as it has been rephrased, "you cannot do just one thing." We have to consider the consequences of our actions as much as we can. Even good actions, taken in isolation, can have tragic consequences. For instance, what if, in setting aside forests in North America and reducing the load on them, we put more pressure on forests in Malaysia, causing them to be cut faster and more disastrously? What is the solution then? Social equity with others? Voluntary simplicity? Global laws?

We seem to be confused between luxuries and necessities. There may be enough forests for necessities, but not for luxuries. We also act confused by the distinction between temporary good and durable goods (nothing is permanent); temporary good are things like cars, entertainment, guns, and drugs (any kind), while durable goods are things like reforested areas, organic farms, well-designed roads, and healthy buildings.

We have spent most of our emotional infancy fighting nature. Up until the seventeenth century Europeans regarded untamed nature as a vast, hostile desert. Wild nature still remains unwelcome in the private garden. We might understand the failure of cities and walls in locking out the forest, or nature. After the Sumerian king Gilgamesh killed the great spirit of the forest, Humbaba, he became possessed with the fear of death and tried to lock out nature with the great wall of Uruk. It did not work then; it will not work for us now—too many people want the wild forest. The forest must be wooed. The forest will haunt us until we give it a new life in the heart of modern culture.

Figure 41. A tree nursery at Altazor Forest in Idaho

Experiment: *Plant One Tree*

In some countries on arbor day, people plant trees. What a wonderful idea. There are myths, e.g. Johnny Appleseed, Seriokai (Avocados), the unknown Indian planter of Banyan trees, and others, of people who spent their lives planting trees. The English forester, Richard St. Barbe Baker, started an international environmental organization—The Men of The Trees (now the International Tree Foundation)—in the 1920s to conserve forests and plant trees, first in Canada than over the planet. He was convinced that agricultural practices, including destroying trees, were ruining the soils. He also worked with the American Civilian Conservation Corps (CCC) to replant redwoods in California. His organizations may have been responsible for over 25 billion trees. He had the original idea of reforesting the Sahara in a military-style campaign with an army of 25 million tree planters.

This is appropriate because this is, or was, a forested planet. On the continents, trees should be the dominant form of vegetation in many ecosystems. Forests are one of the most complex of ecosystems, especially tropical forests. They not only support complex webs of plants and animals, they influence regional wind and water patterns, and even global climate.

We need to replant entire forests now. We need for every person to plant one tree a day for the next two years. By the end of the second year, sequestered carbon could exceed global industrial emissions. Atmospheric concentrations would decline and return to normal possibly within 25 years. However, availability of proper land might be a problem. All fallow arable land (suitable for forests) would be needed. Planting would not be a static goal. Forest ecosystems release more carbon at times. Our uses would change; some plantings would not succeed. 'Dedesertification' itself would affect climate and hydrological cycles.

There are examples of traditional societies, whose agricultural practices increased soil conservation. The Lakandon milpas of the Amazonia, for instance, where the planting of trees built *terra preta* soils, which store large amounts of carbon. Agroforestry trees and shrubs held crops and livestock. Canopy closure had been achieved in 3 years.

In other forests, bats and birds brought seeds of shade-tolerant trees. In southern Belize, agroforestry tripled food production. Using the best management practices, by 2030 up to 6 gigatons of CO_2 could be sequestered per year, equal to global emissions from all of agriculture.

If, at the same time, agriculture was diversified and made permanent, less carbon would be released. Bringing forests back to health over their former extents, and bringing agriculture into an ecological context, are good business ideas. Health is part of any profit. Our personal health would improve also from regular work, such as planting trees.

Experiment: *Should the Infinite Value of Water be Preserved?*
Water—the Big picture: We must hold water use to levels of renewal. We must reuse water internally in a system. And, we must reduce irrigation by 90 percent or more.

Water is the bloodstream of the biosphere. It is not a technical issue first. It is a moral issue. Water has been a crucial component of civilization and urbanization, for drinking, bathing, washing, cooling, flushing, cleaning, manufacturing, and other uses. One of the first technological engines was a water driving engine, which eventually led to the massive irrigation necessary to keep agriculture and cities viable in dry climates.

As a feature of weather and ocean systems, water is also destructive. It can carry away houses as well as dirt. Much money and time have to be spent on its control, with channels, dikes and river barriers.

But, clean water is in short supply. Over a billion people have no access to an organized water supply or to safe sanitation. Unsustainable withdrawals for industrial and agricultural uses are depleting rivers and aquifers. Urbanization increases pressures on water movements.

We are not sufficiently alarmed yet. We need to be. Pollution continues. Nutrients leak from agricultural land, forming massive algal mats on coasts and lakes, killing local vegetation.

Water is simply not valued highly enough! We have found water on earth, but rather than use it wisely and protect its vital cycle, we search for water on asteroids and Mars. We need more knowledge here, and an ecological approach to apply it in our actions.

The water for per capita human requirements to produce food is 1300 cubic meters/person/year. Industrial uses are of a similar scale. This is more than 70 times the 50 liters per person per day for drinking, cooking and bathing. Overuse of water has negative effects on native ecosystems. Vulnerable areas, such as semi-arid tropic savannahs, face large amounts of evaporation. Some use could be partially controlled or avoided.

Water use in food production is 'green' water. Almost 66% of all precipitation is involved in terrestrial biomass production. Only a third (or 'blue' water flow), 40,000 cubic kilometers per year, gets to the ocean. Grey water flow can be calculated, also. Is there red or yellow water flow? Yes! But, it is usually included with 'grey' or ignored.

Water is common denominator. We need to measure its fair, balanced use, then secure both ecosystem and human shares, without exceeding or ruining others. We need a completely integrated system, which would keep native land cover or permanent agricultural cover. We need to create standards. We need to manage land use better, create models with cost-benefit analysis, and force social acceptance of limits and trade-offs, by applying principles of rights and equity, beginning with water.

Experiment: *Why Forests Cannot be Saved*

Regardless of our design sophistication, forests cannot be saved if humans do not stop spreading out (our expansive style of life). Building upward in vertical cities (or arcologies) could save vast areas of land for wilderness, agriculture and recreation. We have the technology for arcologies, but not the will to abandon our spreading edge cities and private lawns.

Forests cannot be saved if there are too many people making demands on them (our unplanned population size). Human nations must be self-sufficient, as much as possible and as soon as possible, which implies fewer people. Smaller populations of people will always be able to harmonize more successfully with wild ecosystems, such as forests. For long-term survival, we need to make optimum-sized human communities dependent on the health of ecosystems.

Forests cannot be saved if humans approach them with inappropriate images (our simplified industrial cosmology). The image of the forest as a machine with interchangeable parts is ridiculous and incomplete. Large industry is inappropriate to deal with local ecosystems, such as forests. We must achieve a mature view of technology, mixing handicrafts, intermediate technology, and even heavy industry (for transportation systems or large building, for instance).

Forests cannot be saved if we cling to our remoteness, watching violence and misery behind a glass wall of television (our insular indulgent psychology). Understanding the ecological relationships in a forest and with other ecosystems requires deep attention and participation.

Forests cannot be saved if humans cannot acknowledge and enforce limits (our four-year, feel-good politics). We know what many limits are, but we are addicted to luxury and growth, so we ignore them. We may have to induce people to adhere to limits with economic incentives, or simply legislate limits to growth and use.

Forests cannot be saved if things distract us from our goals (In *The Meaning of Life*, one of the characters in a Monty Python skit was discoursing on the care of the soul and how to avoid distraction from it, when he was interrupted by a question on sales figures for hats, to which everyone turned their attention. Discussion of forest salvation is like that). Our path to awareness, to forest preservation and right livelihood, is interrupted all the time by trivial concerns. This may be how the world ends.

Figure 44. Preparation for a prescribed burn (Mountain Grove, OR).

Experiment: *How Forests Might be Saved*

The negative reasons seem so convincing, but there are positive reasons. Forests can be saved because human beings are willing to sacrifice their own good for other beings, and because human beings are willing to change to ensure a future for their own offspring. Human beings are capable of learning about, and observing, limits; some of them can acquire wisdom. New images can be forged and perhaps influence a healthier perspective. With forests there is still so much to learn.

Unfortunately, we do not seem to have the will or resources to protect all species in all forests. We can reduce the complexity with models and limit risk to a meager 100 years. Even though we have ways of reducing complexity to a keystone species or to a pyramid of values to understand it, we have not reduced the complexity itself.

When you think about it, there are many keystone species in a forest: Trees, squirrels, millipedes, and mycorrhizal fungi. In a way every species is a key to the whole. Some species, such as mountain lions, can be removed from the forest with little gross change, although there are subtle changes and losses. Any species is an expression of variety, a niche maker—that enriches the ecosystem and expands the habitat for others—and locus of feeling. Any heterotroph, according to Eugene Odum, that consumes autotrophs and excretes matter (with inorganic ions) contributes to the circulation of nutrients and minerals.

Furthermore, any being contributes to the energy flow of the system and is a link in the web. Although an ecosystem can survive without a particular species, it is reduced accordingly. We also use the pyramid model to show importance in an ecosystem. Complex individuals are considered more valuable.

There will always be conflicts between forms of life. So we will have to value lives differently. Complexity can be related to value; this is why we can kill millions of mosquitoes to protect hundreds of humans. A healthy biotic pyramid could be used as a basis for calculating the value of living organisms. Living organisms cannot all have absolute values—even humans. Although this scale may be marginally accurate for the value of beings in a living system, it would cause problems at the human end. However, the forest habitat is more complex than a human habitat.

When we invert the pyramid, the least valuable species becomes humanity, followed by mountain lions and bears. The most valuable are fungi, bacteria and archaea. That does not mean that we should not care about bears and humans (or about fungi and bacteria, depending on how the pyramid turns). Forestry should be concerned with saving the bottom of the pyramid as well as the top, regardless of how it stands.

There have been many suggested practices and principles that foresters can follow to remove some materials from forests and protect the functioning of the forest. Design, for instance, combined with conservation biology and other disciplines, can suggest patterns of cutting that reduce fragmentation, windthrow, and still appeal to people's aesthetic sense. The history of forests records vast changes, migrations, and shifts of forest species; knowing that, we can effect the direction of change by cutting certain species, such as grand fir in a Ponderosa pine zone. Experiences with clearcutting can guide the intensity and scale of cutting on Northwest sites.

We can distill two general principles that people should keep in mind as they work: (1) Learn as much as you can about the history and ecology of the forest you are working in, especially about archaic practices and the previous use; and (2) be humble and cautious about cutting trees—when you take a tree, use everything you take and make sure the rest gets returned to the forest floor. In addition, here are just two of the things we need to get to manage wisely, according to David Perry: (1) Accurate measurements of trophic level productivities, and (2) greater understanding of the complex relationships among plants and soil organisms and how they influence successional dynamics.

A tree, like a pine marten or bark beetle, has biological wisdom; it is clearly self-making, self-governing, and self-choosing (within its limits of treeness, e.g., it cannot walk over the ridge to get better sun). Forests are the books of biological wisdoms. Through our cutting and decision-making and overuse, we are tearing up the books, literally leaf by leaf. Rather than telling the forests what to do, rather than controlling their growth, we need to watch forests to see what they do (this used to be the function of natural history), and we need to let them do it (this requires patience and temperance). Abraham Maslow regards this attitude as "taoistic," and the way to forest health is letting the forest do most of the choosing and working.

Being taoistic means being nonabstracting, nonimproving, nonintrusive, and noninterfering. Maslow also pointed out that most of the action is at the growing tip of a plant. Ecoforestry is the growing tip of forest care. To be a good forester, you almost have to be part of a good culture, and that culture must almost certainly have a good image of the world. Otherwise, the culture may destroy whatever good work you do. A forester cannot just care for the forest. She must also participate in society, and if it is blinded by greed and bad ideas, she must try to influence it to the good.

Figure 46. Restored
Woodford Creek, OR

Experiment: *Is It Meaningful to Design the Amazon Watershed?*
Rivers are freshwater ecosystems that drain 90% of the landmass of the planet; they play a critical part in regional and global biogeochemical cycles, as they transport materials to the ocean. The health of rivers around the planet is declining, to a large extent from human activities, such as large-scale alterations in water and materials fluxes, which makes the rivers less valuable to people.

The Amazon River, with twenty percent of all global freshwater in its flow, drains the largest tropical rainforest, home to ten percent of the species in the planet. The cyclical replenishment of nutrients carried in by the annual floods of the Amazon River Basin creates an ecologically rich area unique on the planet. The annual addition of dust from Africa is also important. The topography is as diverse as the continent. The health of this river, which is threatened by deforestation, drought, hydroelectric dams, resource extraction, and industrial transportation on waterways, is connected to the health of the global system. Clearly, the river has to be addressed on a regional or global scale. Local and national actions have to be coordinated at a regional level. The 30 million people, some of whom live in 350 indigenous groups and the rest of whom are housed in urban areas, depend on Amazonian ecosystems for food, shelter and livelihoods.

The key to conserving the Amazon is its size. A substantial percentage of the forest, possibly over 70 percent needs to be permitted to exist as wild. This is required to maintain the volume of rain needed by the river. The vast stores of carbon in the forests need to be kept sequestered. Development has to be limited by the characteristics and properties of the river. The direct and indirect impacts from large-scale human activities have to be minimized. Keeping the Amazon River functional is important for keeping the forest complete and functional; it may require rethinking the process or industrialization and urbanization in that region.

No application of these forms of rethinking would work without changes in current land policies, which encourage and reward wasteful, short-term exploitation. New policies would have to reward self-sufficient, local ownership and ecological techniques. The land and animals would have to be conserved. Land policies would have to be reformed, with an end to special subsidies. Subsidies granted to large landowners, as well as to mining, sawmills, road construction, and logging, have to end. Laws would have to be enforced. Sustainable products would have to be planted and sold. Tourism would have to be ecological or severely limited.

Ecosystem conversion would be taxed. Land policies would have to be reformed. Under Brazilian law, much of the Amazon is essentially an open access resource so there little incentive for squatters, farmers, or developers to use forest lands or resources in a sustainable manner. To simply clear some land, then move on to another area when the land is no longer viable, is not a viable strategy. Developers can also acquire rights to unoccupied forestland simply by "using" it for at least one year and a day—typically by burning the native forest and establishing some cattle on the land.

To remedy this wasteful use of land, lawmakers in Brazil should consider laws that restrict these practices. Or maybe lawmakers could enforce some of the existing laws like the 1996 law that forbade Amazon landowners from cutting more than 20% of the forest on their land. For whatever reason, the laws on the books have not been effective— deforestation has increased dramatically in the past couple of years.

Deforestation could be slowed or halted with funds generated under a US cap-and-trade or a broader UN scheme to reduce greenhouse gas emissions from deforestation and degradation (REDD, as reported recently in the journal *Science*). The rate of destruction means that the window of opportunity is short and growing shorter. Amazonian deforestation within a decade would mean dramatic changes in climate and productivity. Brazil accounted for roughly half of tropical deforestation between 2000 and 2005. Instead of advocating a market-based approach to REDD, where credits generated from forest conservation would be traded, Brazil expects to benefit from a fund financed with donations from industrialized nations, even though contributors would not be eligible for carbon credits that could be used to meet emission reduction obligations under a binding climate treaty.

For almost 50 years, environmentalists have been voicing concern over the vanishing Amazon rainforest, but their voices and efforts have not been effective at slowing forest loss. Even later, with the popular support of celebrities and millions of informed citizens, with millions of dollars of funding, and the establishment of millions of acres of protected areas for tribal peoples and forests, deforestation rates have increased. Average annual deforestation rates have peaked at 28,488 square miles of forest loss between 2002 and 2004. As land prices appreciated, cattle ranching and industrial soy farms expanded, with new infrastructure. Development pressure on the Amazon is accelerating faster than conservation efforts. We have to design a pattern where limited exploitation is allowed, by tribal groups and industrial ones, and where wild areas are connected by corridors, so most species and their diversity can continue.

Market measures, which value forests for the ecosystem services they provide as well as reward developers for environmental performance, may alter the trend and save some areas from destruction. But, the real drivers of a permanent solution will have to be social and cultural changes in values and ownership. Protection and restoration of the Amazon has to be for the entire watershed. Design and basin-scale management needs international investment and cooperation.

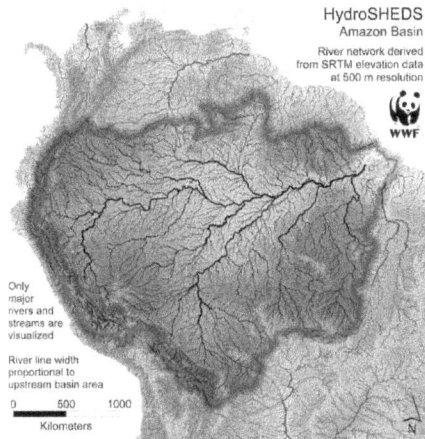

Figure 48. Hydrological Plan of Amazon Basin
(Credit: World Wildlife Fund)

Experiment: *Could We Revegetate the Edges of the Sahara?*

The Sahara desert is the largest hot desert on the planet, with an area of 3.3 million square miles, roughly 10% of the continent, typified by erratic dunes (ergs), plains of rugged rocks (hamada), and relatively high mountains (11,000 feet). The desert is bounded by the Atlantic Ocean, Mediterranean Sea, Red Sea, and the Sahel, which is a semi-arid tropical savanna. About 5000 years ago the Sahara shifted from grassland to a desert. The climate of the Sahara is hot and extremely dry. Plants and Animals of the Sahara are limited by high temperatures and arid conditions.

Since the Romans plundered North Africa for its resources, some people have wondered if the region could be made productive again. Richard St. Barbe Baker started an international environmental organization in the 1920s to conserve forests and plant trees. Baker crossed the Sahara on a 'groundbreaking' ecological survey. He had the idea of reforesting the Sahara in a military-style campaign with an army of 25 million tree planters. He traveled to visit every Saharan leader to promote the project.

More recently, Charlie Paton, Michael Pawlyn, and Bill Watts created a Sahara Forest Project to turn the Sahara into a source for food, water, and energy, as a partial solution to resource problems. The project would combine concentrated solar power and seawater greenhouses in a massive holistic, synergetic 'biomachine' that would turn the region into a flourishing oasis that could export energy to the rest of Africa and Europe. The two proven technologies, seawater greenhouses and concentrated solar power technologies, might work there. Seawater greenhouses could turn saltwater into freshwater for growing crops, even producing a surplus. Concentrated solar energy would use water for steam for turbines. The greenhouses would grow biofuel crops and food. The extra water could help reforest the immediate area and sequester carbon. This design might accelerate the pace of change. The development team expects that the beginning Sahara Forest Project would span 20 hectares and cost about 80 million Euros. However, the consequences there and 'downwind' have not been adequately considered.

In another attempt, the southern side of the Atlas Mountains in Algeria's desert is being turned green, as a vision of the civil engineer Madjid Abdellaziz. An area has been turned into a wide green oasis, where now fruit trees, vegetables, wheat, and even truffles grow in the desert sands.

And, there are 11 nations participating in 'The Great Green Wall' tree-planting project to attempt to halt the spread of the Sahara desert. This project aims to plant a wall of trees 4350 miles (7000 km) long and 10 miles (15 km) wide across the southern edge of the Sahara desert.

Geoengineering to terraform the entire region has been suggested as a corrective for global warming. Leonard Ornstein, David Rind and Igor Aleinov propose a plan to pump desalinated seawater from the coast to irrigate fields of nonnative Eucalyptus trees, which would replenish the soil and cause more rainfall, allowing for more growth, fixing tons of atmospheric carbon, equivalent to the total emissions of the planet today.

Craig Embleton addressed the potential for reducing atmospheric carbon dioxide levels through solar-desalinated irrigated vegetation of the Sahara and Arabian deserts. Many reasons are listed for greening the Sahara, including carbon sequestration, restoring its forests and grasslands with their complements of wonderful species, and feeding a growing human population. One reason is the human idea that land should not be wasted but be properly exploited. Certainly a green region the size of the Sahara would be an impressive carbon sink, but the effects of this change would result in many other unintended changes, many of which could be negative, such as a shift in westerly winds. The scale of these proposals would result in many unforeseen or unintended problems.

If we tried to return the Sahara to an earlier state, how would we decide which state? The pre-Ice Age state, the Ice-age state, the first post-ice-age state, the current state, or a totally anthropogenic once?

Any large-scale change in the Sahara would bring changes to local weather patterns. Increased local precipitation could certainly promote more growth in plants, both desired and weedy species. Furthermore, insect pests, such as locusts, or other plants, such as molds, would also increase, threatening the target species or the balance of the system.

Regardless of how desirable these might be, it is likely that they will influence or change regional and global patterns. A newly forested Sahara might create a low-pressure atmosphere that would draw more moisture from central Africa, perhaps drying out the tropical forests there, or it may bring more polar air to Europe. The wind-born dust that is carried from the Sahara to the tropical forests of South America would decrease significantly, perhaps reducing soil fertility in those forests, as well as reducing the quantity of Atlantic Ocean life. Similarly, the moisture that would have been picked up by the dust would be lost to the Amazon forests.

With a system of this scale, we might reach the limits to knowledge quite quickly. We have not created an Encyclopedia (or Wikipedia) of information about species interactions or about large-scale ecological changes. Data on desert soils are low. Data on carbon sequestration rates and transpiration efficiency is low or missing. Given the real limits of ecological knowledge, and given the certain problems with large-scale projects, we should limit the restoration of grasslands or forests to many small local experiments, beginning with those already started and expanding to sheltered areas, especially at the northern and southern edges of the desert to keep it from expanding into the Sahel.

The deep history of the Sahara and the planet suggests that deserts occur under certain conditions that balance other areas of the planet. Many scholars suggest that the planet cannot maximize the number of efficient, productive systems at any time, and so some systems dry out. Others suggest that maximization might not be desirable as a characteristic of dynamic systems existing in a nested complex of changing conditions. It might be possible that a maximal system is vulnerable to a more complete collapse, and that the current ranges of systems is more stable in the long run.

While it might be possible to reforest or regrass the Sahara for our benefit, the project could force other changes, such as drying out or damages to agricultural systems. Reducing the amount of anthropogenic carbon dioxide in the atmosphere is a good goal to keep the planet cooler. New vegetation would sequester more carbon, especially in longer-lived organisms, such as trees. New vegetation would change wind and moisture patterns, as well as albedo, which could increase the local temperature.

Land-use changes, as a form of global geoengineering, might be preferred to the technological production of orbiting mirrors or ships full of iron-filings, but the scale is still a problem. The scale would have scale effects on other neighboring ecosystems. Monitoring would be necessary.

Irrigation itself might be minimized in the desert. Irrigating with seawater might give an initial boost to ecosystems, but they would be saltwater wetland systems and not fresh.

The green wall should be tried to help the Sahel, but with the understanding that the wall is permeable. Sand, as well as air, pollen, and animals, will slip through. Forest restoration should be attempted only with native African species, such as acacia and cedar, which have adaptive interactions with other endemic plants and animals, and not with fast growing exotics that require a fire ecosystem.

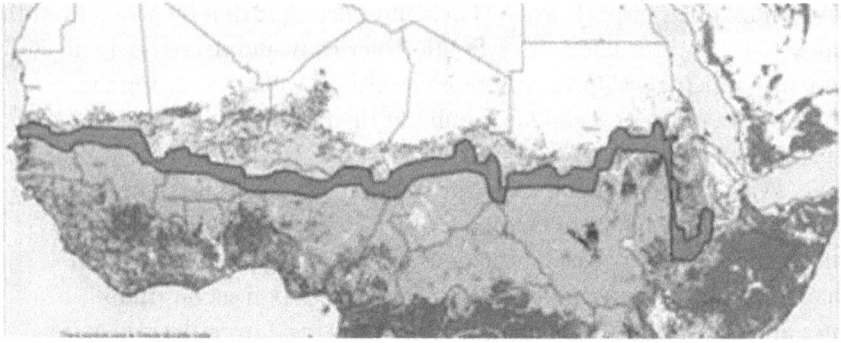

Figure 51. *Map of The Great Green Wall project in Africa to halt the spread of the Sahara Desert.*

Experiment: *Why are Trees Tall? (How to Reduce Diversity)*
Green bacteria, which learned to store solar energy in sugars, are not tall. In fact they are really short, especially floating on the surface of water. A billion years ago, bacteria were seduced into plant cells. We call them chloroplasts now, but they still reproduce like bacteria. Trees depend on chloroplasts to store energy. Over evolutionary time, trees became taller. Even alone, among grasses, a tree will grow tall. It's a growth habit now, written in the code of trees.

Trees normally grow in large groups. In fact, growing in groups leads to competition to be higher than the others, to collect the photon rain as it hits the earth from the sun. Growing tall, however, requires a significant investment. Most all trees make that investment to be part of the high rolling meadow on top of the forest. This uniform height of each species results from competition between trees. Each tree has to invest energy in wood for support itself at a good height.

The height limit of trees is where the cost of growing another inch is greater than the gain of extra photons by growing. Trees grow as tall as they can to compete, but not so tall that the cost of being another inch taller would be prohibitive. Trees grow to increase their individual benefits. The forest evolved as a collection of individual trees competing.

An economist would say that investment is wasted. If all trees were an inch tall, that energy could go into horizontal spread or needle or leaf production. The tree trunks are monuments to the costs of competition. We know that there is cooperation in nature as often as competition. On the other hand, for another species, humans, modified tree trunks have value holding up human houses. If the forest were a planned economy, trees would be limited in height to an inch or two, just like bacteria in the ocean are limited to inches.

An economically-minded botanist could plan and rationally

manufacture a short forest. It would be more efficient as a forest with its energy use. In a planned rational economy, the forest would minimize costs and maximize benefits. Most animals would also be small. And slow. No reason to be large to eat poorer food. All animals would have good food. Predators and prey would not have to be fast or faster. Think of the savings in energy! No more escalation of the evolutionary arms race. Maybe there would be less need for camouflage?

In the planned forest there would be less need for extravagant size or speed. In fact, there would be no need for diversity, for any differences at all. In fact, all predation of chloroplasts would be eliminated; there would be no predators, no prey, so no colors. No insects at all. Pollination by the wind only. No birds to eat insects. No bird song. No mating, so no music, either. Nothing beyond straight production. No running or hiding, no fighting or competing. Just production. Production for human consumption. No animal skins for clothing. All clothing could be grey. If mating was as planned as forests and food, maybe no clothes would be necessary. In perfect social equality, no large houses or apartments would be necessary, just basic boxes.

In fact, all body plan differences would be unnecessary. All large bodies would be unnecessary, for domination or prestige. And now, we are back to bacteria, green bacteria. Nothing else is needed. And now intelligence, which costs so much, is no longer needed. Perfect economy. Perfect world.

Figure 53. Forest Floor
(Altazor Forest, ID 1999)

Part 3: The Wild

Experiment: *Charting Northern Hemisphere Wolf Paths*

The great gray wolf once inhabited a wide variety of habitats throughout most of the northern hemisphere north of the 20°N latitude. Because the animal preyed on livestock and competed with humans for wild prey, it was extirpated from much of its range outside of wilderness areas. Environmental awareness in the late 1960s brought for the wolf legal protection, increased research, and favorable media coverage. The species has increased in both Europe and North America, and is beginning to reoccupy semi-wild areas and agricultural land, where it is blamed for causing damage to livestock. Because of the wolf's high reproductive rate and long dispersal tendencies, the animal can recolonize many more areas. In most such areas control will be necessary, but the same public sentiments that promoted wolf recovery reject control. If wolf advocates could accept control by the public rather than by the government, wolves could live in far more places.

There is reason for caution in the design of any reserves for wolves. Data on species like the wolf are almost never complete. Not enough ecosystem details are known. Human modification and destruction of habitats has a decided effect on the rates of gene flow between populations. The reintroduction of species from conservation centers will be meaningless without great and successful efforts to restore the patterns and depths of habitats. Although genetic information is essential, knowledge of large-scale patterns is crucial. The structure of a community, from behavior to culture, in the context of a changing environment, is what determines the size of a wild population.

It is possible to set aside large areas for the wolf in the northern hemisphere and some natural corridors between the areas, especially mountainous areas. Wolves had no significant problems dispersing through rural areas in Eastern Europe or western North America. It would also be possible to link almost every area with new corridors. New links would not require much coverage and could be relatively narrow. This should have little effect on the species and populations, since 13,000 years ago most of the hemisphere was available to wolves.

Most of the wolf paths would be in lightly populated areas. Where human density increased, the corridors would have to be marked well and monitored to make sure that they stayed open. Where they crossed human corridors, large bridges or underpasses would have to be built. Where the corridors were constricted by geography or human settlements, then long bridges or rock-walled lanes would have to be constructed. The impact on human constructs should not be great, however.

Finally, restoration is phenomenally expensive, conservation is very expensive, but setting aside preserves or paths is only moderately expensive, and much of that resides in perceived losses from the denial of exploitation. To set aside a series of wolf paths in the northern hemisphere may cost $50 billion USD in modifications and twice that much in 'lost' revenues from resources. Because wolves seem perfectly comfortable using human roads, trails or railway tracks for travel and for hunting, that number may be reduced. Significant amounts of the money would be required for education and signage. Within the limits of altruism—restoring what we have been destroying—we simply should set aside 54% of the planet as reserves and let evolution proceed, with the self-organizing configurations that we term species, genomes embedded in ecological communities (Wittbecker, 1976).

Experiment: *Are We Dependent On The Wild?*

We are actors in a tremendous presentation. We are players in the evolutionary play in the ecological theater. We are also the audience. This metaphor first formed the basis of a worldview where nature was a theater of violent competition. The frame of the metaphor carried the conceptual baggage of the culture in which it originated. So, over time the play supported the idea of superiority of "favoured" races in the struggle for existence and emphasized the role of competition in biological and cultural situations, at the expense of other interactions.

Alas, the metaphor needs to be expanded. All species play a temporary role in the local stage in the ecological theater. Herb Hammond notes that all the actors and acts are essential. We are foolish to think of some species as more important than others—that is ignorance or wishful thinking. All species and things contribute to the functioning of the whole, including rocks and gaseous elements. But, some are invisible to us because of their size or longevity. Some play their roles in a clump of soil, others in continental landscapes. Some acts last less than a second; others take millions of years. Even if actors seem to leave the stage or the act seems over, they influence the play with their corpses and elements.

There are many stages playing simultaneously in the theater, and it is not a one-act play. The human play has converted some of the stages, then subverted and subsumed others. The human actors are ridiculously egotistical and ignorant, pretending that the stages and theater is for them only, and other are support characters. They pretend that the less important people are in the audience, but the only audience is other stages with partial perspectives. Each is audience for the other.

And, some plays are embedded in others. One basic play is the evolutionary play, where autopoetic beings drift through filters in the mor-phogenetic landscape. Another play is the human conversion of ecosystems and the urbanization of human communities (with some companions, familiars and pests).

We have trouble understanding the theater or the plays because of our physical, temporal, psychological, and cultural limits. We see species and ecosystems as individuals, when most of them are in fact communities. We see walls and barriers, and not permeable filters. We see philosophical constructs of classes in isolated locales. We have problems dealing with motion, indeterminacy, ambiguity, and vagueness. We are seduced by logic and fallacies to believe that we can understand and control the play.

Although the scales involved in a planetary theater are linked by processes, in fact the scale of the planet and its evolving life forms, is significantly larger and longer. As important as human changes are to us, and to the ecosystems from which we emerged and on which we depend, for the planet those changes may only bump the global system to another stable state, well within the range of those from the past 2 billion years. We are the stewards only of ourselves and our companion forms, not of the planet. It is the whole that is wild, and we depend on it.

Experiment: Can We Avoid the Return of Silence?

Can we avoid another Silent Spring? Once in the 1950s, we used poisons to kill pests like mosquitoes or aphids. DDT and other organochlorine insecticides also killed large number of birds, from wrens to eagles; it ruined the egg shells and bones.

These are now banned. Other insecticides and pesticides replaced DDT and it cousins. So, birds have been making a comeback, until now.

Now birds, especially farmland birds at first, are declining again. It seems that the new 'safe' pesticides, more often neonicotinoids, are wiping out the insects, including honey bees, which are the food for birds. Again the effects are cascading effects that might have been predicted. These artificial solutions always overkill or kill all species.

They also have side-effects that can influence human health and genetics. Many children and adults are now experiencing unusual forms of cancer (beyond the ones that are age-related). This raises a question. How can we raise food without destroying ourselves and everything else?

We might use our knowledge of genetics to engineer a spray that would target only corn borers or wheat rust, but that might spill over to other populations or have unforeseen consequences. We might create foods that would repel all pests, but would be mostly inedible fiber, hard to chew or digest.

We could simply switch to organic agriculture or to other complete systems such as Permaculture or M. Fukuoka's Natural Agriculture, where crops are interplanted to varying degrees. In such a system, nature would be busy and noisy.

If we want to avoid the deadly forms of silence, and if we want keep hearing the joy of birdsong, we will have to be more careful with our simplistic use of monotonic biocides or perhaps do without them. We may have to begin conservation or breeding programs to being back many of the species of birds, especially the most colorful and harmonious of them.

Experiment: *Can We Successfully Woo the Wild?*

Rene Dubos takes the heroic love-adventure of humankind to be the wooing of the Earth. The wooing of the earth is both sweet and sour— sweet because humans can create enchantment within nature, and sour because they can spoil desirable places—possibly to the point of ruining nature's recovery mechanisms. The wooing of earth implies more than converting wilderness into humanized environments. It also means preserving natural environments in which to experience mysteries transcending daily life and to recapture an awareness of the natural forces that have shaped humanity and a wild earth.

Dubos celebrates the human ability to woo the earth by creating new environments that are "ecologically sound, aesthetically satisfying, economically rewarding, and favorable to the continued growth of civilization." While that is noble and just, there are two erroneous assumptions contained therein. First, civilization does not have to grow to develop; it is development that is important, not growth, which is plainly unfeasible. Second, "to woo" means to solicit love or to make love. Dubos assumes we will remake nature in our image. And love, as everyone should know, entails respect or reverence. It does not remake the loved one in one's image; it respects and allows freedom and difference. That is the larger meaning. We should "improve" on nature where we live, but we do not need to try to improve on all of nature. We do not need to domesticate all beings for human use. Joseph Wood Krutch recognized that ecology without reverence or love is only shrewder exploitation.

That does not mean that we cannot change places by living there. We can create places only by living there, by slowly adjusting each to the other. Human modifications of the earth can be lastingly successful only if their effects are adapted to the invariants of human and physical nature. And, this is the creation of valuable ecological capital. Ecological management can be effective only if it takes into consideration the visceral and spiritual values that link us to the earth. Although some of nature can be regarded as a garden to be cultivated, large areas should be preserved as untouched re- serves or necessary systems for global balance. Other areas could be studied or urbanized as population centers. Our tendency to completely redesign nature, rather than tolerate and cherish, is dangerous; it leads to "curing" abnormal people and "solving" weed problems. We do not respect the wild, complex side of human nature or wildness in nature.

A true definition of wilderness is not found in human needs and desires. Wilderness is not for people. It is for itself. It makes and defines itself. For humans, it is a land classification. But a unique one, for it is one that is nonhuman and unmanaged, unknown and unused. Wilderness is not incoherent until human beings give it form. It already has form. Wilderness is made by the life-images of the species and beings that inhabit and inform it. Wilderness is not empty. It is full with living beings whose goal is fulfillment in living.

The primary justification for wilderness is as a sanctuary for other beings that constitute the richness of the earth. Many of these cannot live with humanity; they are allopatric and anthropofugic (actively fleeing human presence). If, as many believe, wilderness is necessary for human development, then its protection can be justified on that basis also. But, its value to humans consists of its separateness, as well as its diversity, aesthetic nature, its economic benefits in recycling wastes and trapping energy, and for recreation, inspiration and teaching.

Many voices defend wilderness itself. But more urge its use. We humans are ambivalent. We want areas that bear no record of us, yet we want to leave our mark everywhere. We want nonhuman places, but we want to live everywhere. We want to use everything, but save it. There are many alternatives, from indifference or conservation, to using incentives.

Humans need a human way of being in human worlds. But, these worlds are derived from a wild universe, which we need to live fully. When we understand our roles and relationships in nature, then we will not be managers or stewards, but participants and sharers in experience. An ecological basis gives human activity dignity. But that in no way negates the life-images of nonhuman beings, which are all centers of experience.

We must emphasizes the value of all nonhuman beings and try to use as few individuals as possible from as many species as possible— moderate yields, within the limits of the food chain. Many species should not be used at all, whales and wolves, for instance. Plants and animals are as much of our heritage as art, history and tools. Ignorance of one is just as sad as ignorance of another. The survival of society depends on an expanded ecological consciousness. The spirit of humanity depends on this consciousness to place us in a proper relation to the wild places of the earth, taking what it needs, but letting the rest be.

Perhaps new technologies and new philosophies may permit larger human populations or the expansion to other planets or solar systems, but we must be cautious with what wilderness exists here. Conservation and creation must be tempered with preservation. We are too ignorant to tamper with everything. If we cannot live sensibly on earth, the emptiness of space and wilderness of stars will never be home.

Experiment: *Should We Keep the Wild Apart?*

By the 20th century, appreciation of wilderness came from those who realized that their lives had been impoverished by being set apart from areas where humans did not dominate. At the same time, now that humanity was competing for protein in every ecosystem, wilderness was regarded as a resource. Most scientists since Bacon have considered all of nature as a resource for the creation of human wealth. Theodore Roosevelt helped promote wildlife biology with new parks and the new field of wildlife management (an oxymoron). Later, Aldo Leopold and Rachel Carson cited the Kaibab plateau mule deer as lesson in why predators are necessary.

The solution to the quandary of wildness is: Keep it apart. Paul Shepard argues that 75% of the land area should be left wild in a techno-cynegetic society. Reserves of every kind of region need be saved. Shepard concludes that the consequences of ecological conflict in technological, humanistic civilization are either exile or sanctuary; and sanctuary is the only solution available to the humanitarian ideology. The idea of sanctuary recognizes the multiplicity of factors necessary for viable populations. But, although there are sanctuaries for frogs and ferrets, it would be impossible to establish one for every creature. Shepard discounts the humanitarian objective as considering only "worthy" species. Yet, if all species were considered, the whole planet would end up as a sanctuary. He considers sanctuary an unfeasible 19th century political solution, based on a time when space was unlimited. Shepard states that exile (or extirpation or extinction) and sanctuary are allopatric choices (meaning "life not occurring together"). Allopatry is seen as being consistent with the tradition of personal property, domination of nature, and model of the nation state.

What Shepard overlooked is that humanity was never sympatric. It never lived together (the meaning of sympatry) with major carnivores. Therefore, allopatry is consistent with the nonexploitation and nondomi-nation of nature as well. Sanctuary as personal property or as nonhuman property is a moot point if habitats are saved from destruction. Domestica-tion and enslavement are sympatric forms. A conservation program, despite Shepard's argument, cannot be based solely on sympatry.

Not all species occur together in the same place. Sympatry describes only those that do. At the habitat level, allopatry is the most intelligent use of available resources by animal communities. Large herbivores, such as elephants and rhinoceros, may choose poorer quality food and avoid com-petition with smaller animals and exploit an untapped food source. Many interactions between different species contribute to the mutual benefit of the members of the community, as well as to the community itself. Human-ity must be allopatric with most wild species and allow them to develop independently in their own places.

Separate areas are necessary. Gary Snyder questions whether

complete compartmentalization is healthy. He suggests that some land is saved like a virgin priestess, while other land is overworked like a wife, and some is brutally reshaped like a girl declared promiscuous. This argument is certainly good for human lands, but wilderness lands are being compartmentalized for the use of other species (which do not share our human treatment patterns of females, for instance).

Saving wilderness means saving large areas of land. It means placing wilderness, which is support for cultivated and industrial areas as well as for neopoetic communities, off limits to development and perhaps to any use. Every community should be allotted a place for self-development, a foundation area. Wilderness is not land that is locked up, as several U.S. senators have claimed; it is quite interactive. We are already living with it. Foundation areas are not museum pieces, as Charles Birch seems to think. They are functioning ecosystems. Only humanity is directly absent.

Wilderness costs are sometimes considered too tremendous to bear, in lost opportunities, resource use, and administrative overhead, according to C. Lehmann and others, but that is without considering the free services of wilderness. That conclusion is based on the assumption that everything that could be used should be used. Wilderness areas, left by themselves, would require a minimum of management costs. Often, when it is suggested that wilderness areas be left without human management, critics conclude that humanity is not being considered as part of the environment. This is an unfortunate conclusion and untrue. Humanity does not have to convert every system to be a part of it. We simply participate indirectly.

A high human civilization has to limit its consumption of natural resources and its place in wilderness, to account for wilderness and future generations. Wilderness is a limiting factor in the health of human civilization. It is probably not a precise function, but we may be close to the minimum for the earth. As Gregory Bateson pointed out, it is not safe to be limited by lethal variables.

There is a more basic reason to save wilderness: It exists apart from human concern and industry, and it is the only sanctuary for wildlife apart from humanity. Wilderness is filled; it has its own values and uses, even if they are nonhuman ones. It is wilderness itself (*Vastitas ipsa*). By definition, it is not manageable and does not need to be managed.

Figure 60. Berry-drunk woodpecker

Experiment: *Can We Manage Wildlife Trade & Consumption?*

People love animals, and want to benefit from their grace or power. Unfortunately, they believe the only way to do that is to kill the animal and grind the organs into small dry potions and eat them. Few people have ever seen that work, but nobody wants to give up superstitious behavior if there is any probability of success. So, in a population of 7.3 billion people, a large number are trying to kill every charismatic large animal so their own penis, wisdom or courage might be regarded with awe.

Animal teeth and bones have been used traditionally by artists. One problem is always how to handle sales of art objects made from animals before protection laws were in place. These historical art works might distort the entire art market if no more ivory, horn, or bones were ever again used. But, it is no different than the mortality of artists limiting sales.

Illegal wildlife trade is high financially, as high as all trade in all drugs. Because poverty is rampant in the Americas, in Asia and Africa, there are many people willing to risk the penalties for killing these animals, which are also symbols of the human nations that overlay their lands and territories. In their natural habitats, they also draw tourists from around the world, and that money income often exceeds national industrial output or drug sales.

Rationality does not, cannot compete with high, easy income. Nor can moral restrictions or outrage. There has to be a way to protect animals. One suggestion is to enforce the death penalty for poachers; this could be done immediately as long as there was one neutral witness. Financial incentives for rangers, who sometimes are the worst offenders of all, might help them to work on behalf of the parks or animals. However, some adventurers would still risk that to get rich selling the ivory or tiger penis. And, the trend may continue until some species, such as lions, tigers, rhinoceros, and elephants, are pushed to extinction.

What would happen if every nation or the United Nations made the entire trade in wild animals, and their magic parts, legal? Killers (or euphemistically collectors) would be licensed, with continued high penalties for poachers. Licenses would be limited. The number of takes per species would also be severely limited, depending on its population characteristics and ability to tolerate a certain number of deaths beyond those from natural causes. The entire process, from identifying an animal, through killing and reduction to pieces, and sales, would be monitored closely. It could also be discontinued until the population recovered to a scientifically-determined minimum level. Finally, the animals pieces, and possibly the value added by the artist, would be taxed at very high rates. That might shift purchases to the very rich, but that is not much different from now. And, the prime animals and species would be protected more than now.

And, if it did not work, we could enforce a total ban on all killing and sales; furthermore, illegal materials could not be inherited (a state take).

Experiment: *Should Some Of Everything Be Saved?*

We spent 24 years restoring a 70-acre site that had been logged in the
1930s through the 50s (and bought by us in the 1960s, although no bank
would lend on it). The loggers did leave a few legacy trees and seed trees
when they cut. They also established three mills on the property; when my
neighbor and I decided to start a tree farm (both properties) and mill (on his
property), we were able to salvage enough parts to build our own sawmill,
which was powered by a steam tractor on loan for two years. The slabwood
piles and sawdust pits were also mined for firewood; the sawdust was burned
in a Belgian stove or used for kitty litter and then composted.

I kept inventorying the plot for regeneration and for rare species,
such as elegant cat's ear or white pine (once a dominant species favored for
match sticks at the turn of the century). In the late 1980s, we decided to
gift the land to The Nature Conservancy, when we realized that they were
trying to preserve a necklace of lands around Moscow Mountain, six miles
north of Moscow Idaho. Our property was about a mile from the university
forest and about a mile from one property the TNC already had.

When the paperwork was drawn up and ready for signing, we went
over it with two people from TNC, one a lawyer, the other the local botanist
who had assessed it. We noticed in the small print that the TNC reserved
the right to sell the property to buy a more valuable property that might
come up for sale. We crossed out that paragraph and initialed it, but the
TNC would not take the property, because they said, "it just is not valuable
enough." We pointed out the rare flowers and 2 white pines, then noted
that other pieces of the necklace, singly, were not any more valuable. The
impasse was not to be resolved. They refused our corrections; we refused to
sign with that clause.

Over the years I have been very disappointed with TNC, mostly due
to their utilitarian approach to land. They have been very successful getting
land and money. They have embraced the concept of the "working landscape," which is becoming the centerpiece of their agenda for conservation.
George Wuerthner cites the TNC's chief scientist as saying that their task
is no longer to "preserve the wild, but to domesticate nature more wisely."
Apparently this is said with full knowledge that domestication has brought
about terrible problems, from landscape conversion and monstrously eroded
landscapes to extinctions and habitat destruction.

A working landscape or working wilderness is just a euphemism
for ruthless and destructive exploitation—just the thing that the TNC, as
well as The Wilderness Society and others, once tried to reverse, with a
careful definition of wilderness that included minimal human presence,
much less exploitation—this is the bad definition of exploitation: Use or
extraction without limits. At a time when people are calculating the services
of wilderness, services that are not provided by domestic landscapes in

any meaningful sense, there is no justification for transforming wilderness into cattle range deserts or clearcut mudflats. This approach minimizes or destroys biological diversity, interrupts evolutionary processes, and reduces the ability of the system to generate clean water and air. Ranges, farms and tree farms are not ecologically benign. The TNC has simply caved in to the ignorant self-aggrandizement that characterizes the "anthropocene" age (anthro-obscene age?).

Wilderness, and predominantly wild landscapes like forests, have direct benefits for the planet, especially in terms of regional and global cycles. They also have incalculable benefits to our humanized landscape that share a bioregion. Without these wild areas, humanity will not be able to provide for its current population, much less an expanding population. If the TNC were to recognize this, I would be pleased to give to them again, maybe a forest or two (the results of random luck and wise investments).

Experiment: *Should We Put Sensors in All?*

It is so simple. We do not know where animals go and hide. We do not know when or where they die or are taken for illegal or immoral purposes. We have hard trouble finding them and studying them, for a variety of purposes: To know about them, to help or preserve them, or to use them more efficiently.

But, now our technology has given us some small, noninvasive, inexpensive tools. We need only implant a sensor into every medium or large animal and then track it by satellite. If the motion stops or suddenly moves to a city or packing house, we can swoop in and rescue it, or avenge it legally. This would stop to a large extent the trade on Tiger penis and rhinoceros horn, as well as other odd, and mostly ineffectual, superstitious medical practices.

Sensors could be used to track everything, including trees and rocks, which are sometimes vulnerable to hijacking for expensive wood or lawn decorations. We need to apply this to everything in nature, trees, rocks, animals, birds, insects—everything. But, this would require many more biologists and veterinarians, as well as technicians and field personnel. So, it would be expensive, but good for those professions, which often have over 50% unemployment rates (believe me, I know first hand). And possibly good for animal and plant populations (those unknowingly in danger).

Of course, moral questions arise. Are we being too invasive? These are wild animals that expect to go about life unpursued by cameras and blowguns, and often biologists are more insistent than predators, and better-equipped technologically. Are we being too controlling, moving some individuals to unfilled niches, or destroying others to balance a population (often restricted in too small a reserve), or trying to mate them to increase some populations. All of our actions may be interpreted by some as cruelty or the bad hands of science.

Perhaps biologists and everyone else involved could take an oath, like physicians or ecoforesters. That might be enough. Perhaps a monitoring committee or an oversight committee would be suitable, following new rules for the proper use of these technologies and for the proper procedures to follow to minimize harm to the subjects, who are not unlike us in wanting to live well and to have secrets.

Figure 64. Placing Sensors in wild animals & plants. (Credit: Google)

Part 4: Climate & Planetary Limits

Experiment: *Does Evolution Favor Us?*

We seem to be the golden child in the pantheon of change and survival. We have not only survived, we have prospered and expanded far beyond the range of any life form, except bacteria (which makes up a good percentage of the human body, also). Most of us not only accept the theory of evolution, we also look to it for indications of how things work. It used to be that evolution resulted from basic competition, and that sorted out the winners. But, more study revealed that the whole spectrum of other interactions, including cooperation and parasitism, shaped species and environments as well. Although the theory of evolution, for instance, is not a good basis for an ethics, its perspective can supply principles on which we could base values:

It is good to remain adapted, within limits

It is essential to encounter the environment to which we are adapted

It is essential not to destroy the environment to which we are adapted

It is possible to modify the environment within limits to help ourselves live better

It is good to cooperate with other species to exploit a reasonable number of them

It is good to understand that species and environments are tightly linked and should not be separated

It is good to be balanced with energy, places, productivity, and other beings

It is good to not interfere with the process of evolution

It is good not to interfere with the regional and global biogeochemical cycles modified by life and necessary to its full complexity

It is good to respect evolution, communities and the planet

In the larger view, evolution is value-free; it does not judge the beauty or values of animals. If we consider numbers, then evolution favors bacteria and ants as much as, or more than, human beings. Creation and destruction, beauty and ugliness, are expressed in one complex pathway.

We may not be selected to survive, if we refuse to work within natural limits. We may succumb to some disease that we have released from a modified ecosystem. We may starve when our food crops cannot resist hungry insects. But, if we follow a few principles and are humble, we may continue to thrive.

Experiment: *Does Religion Force Climate Change?*

What have the desert people done to the desert, their historical home? Did they enshrine the desert as the most desirable of biomes to live in? Has that allowed them to express less concern over the conversion of lush grasslands and forests to deserts. Do desert religions unconsciously long for deserts?

What have those people of forests, seas, grasslands, and highlands, done to their environments? Have people with strong religious beliefs in preserving their territories fared any better? Preserved them? Or allowed them to be made into deserts? It seems that the scale of population, or the extension of agriculture, or the domination of industrial civilization, has proven to be too much for any one religion to take a stand and save their cherished environments.

What have the people who emigrated to a different habitat done to their new home? Try to make it into an image of the old one? Did that do more harm, especially to islands?

Have these conversions added up to climate change? Religions have many limits, even universal ones. Could religions be rehabilitated with an ecological perspective? If not, could a new religion? Perhaps similar to New Age philosophy, but with more power and commitment. What would a new religion look like? What things would be different or added? Would it have an ecological perspective, and a strong devotion to balancing nature? Would that be enough? This is another new thing we could try.

Experiment: *Is Global Warming New?*

Did the megafauna in North America and Australia die out about 13,000 years ago due to global warming, or from stress or overhunting? Or a storm of convergences? In earlier periods, such as the Jurassic and Cretaceous, was it warmer than now? Was that because of global warming? How stable is the atmosphere? Do coral reefs and forests make it more stable? Do religious images or industrial processes make it less stable? Do the characteristics of life cause things to fluctuate more or less? Including human actions? At what point does climate become global warming? Is it a symptom, problem, or nothing? Are green houses gases increasing? Is the average annual temperature increasing? Is it a positive feedback system? Is it too complex for us to understand?

Is global warming bad? For us? For our companions plants and animals? For the Planet? Should we try to control it? Will it reverse on its own and head towards another Ice Age? Can we control green house changes? How could we try? How do we know if we can? Should we try anyway? What kind of behavioral changes could we make? What kind of technical things could we do? Mirrors? Ocean Doping? How would they work? What other effects would they cause? What is the worst that could happen?

The accused villain CO_2 is very long-lived, over 100 years, and 56% of it from industrial processes is still in the atmosphere. CO_2 interacts with water vapor, methane, and sulfur, affecting how those gases retain heat or contribute to cloud formation. CO_2 is good for some plants, mostly trees. But, plants in extra CO_2 have tougher leaves, less nutritional value, and higher amounts of defensive chemicals. One species that will benefit from changes will be parasites. Mosquitoes will spread malaria in some places.

The CO_2 is already in the air. We have no easy way of getting it out. So, the course of climate change is set for next several decades, or 15 thousand years, depending. Significant CO_2 is removed through natural processes, which also generate quantities of CO_2. We could remove quantities, but it would be expensive, and may not be at a scale large enough to be effective. We do not know the threshold for anthropogenic change, but it could only be 2.3 degrees F according to Steven Schneider.

There is no direct, immediate feedback to counter or correct over-shoot. The wastes become global, that is, CO_2 can leave still auto exhausts despite the concentration in the atmosphere. There is feedback, but it is delayed for a long time by the size, flexibility and redundancy of the system. D. Meadows notes that a person who makes a decision based on feedback cannot change the behavior of the system that drove that feedback. Decisions only affect future behavior. There will always be delays, since nothing can react instantaneously. So, warming will continue.

Experiment: *Are Some Global Problems Insurmountable?*

Civilizations have experienced problems that seem insurmountable—and those civilizations subsequently collapsed and disappeared. Many of the problems are structural, logistical, or spiritual. Economic decline can lead to stagnation, disease and collapse. Many have to do with political power distribution, especially if related to the desire to conquer, control, and make uniform other cultures through war (and now the threat of nuclear war with nuclear winter). Over-administration has its own seemingly limitless costs that can lead to collapse. Imbalance as a general condition can lead to col-lapse. Some of these problems could have been solved or have been solved for a time.

Urban problems, and some national and regional problems, seem insurmountable. Some problems, the largest ones such as earthquakes and tsunamis, are the result of global phenomena and seem insurmountable. The problems listed below, having to do with water, heat, crime, and struc-tural maintenance, seem to be insurmountable.

Drought seems insurmountable. Water has been a problem in many civilizations, from Mesopotamian and Indian to Chinese and American. The changes in wind and rainfall patterns, or river beds, have resulted in drought. In Mesopotamia, for instance, cities compensated for declining rainfall by irrigating wheat and barley with canals. But that lead to salt retention in the soils. By 1700 BCE, it was written that the earth was white with salt. After much intensification, the land collapsed. Many of these things are long-term problems and do not become evident for several generations. They are also very difficult to reverse. For a society that needs surpluses to continue, with growing dependents and growing trade, there is little flexibility to change. The only way to avoid the problems was to let the land be fallow for long periods until salt washed out. This alternative was impossible due to food demands.

When cities started to fail, as a result of long or sequential droughts, people were able to emigrate. Many returned to herding, or when possible, hunting and gathering. For thousands of years, starting possibly 11,000 years ago, people participated in a cycle of emigration to rural areas when time were bad and immigration back into new cities when the ecosystems recovered and could be made productive again.

In complex, self-regulating systems very small changes have large consequences. In some cases, where conditions like drought are cyclic, in the Sahel region of Africa, humans expand during the good times, only to perish when the drought returns. In other cases, human activities, such as deforestation or overgrazing of herds, can cause weather changes and contribute to droughts. The scale and rate of changes allows people to view the situation as natural, but once these catastrophes pass a threshold, the people and their cultures have been trapped by their demands, and only severe reduction or collapse can allow the system to regenerate.

Drought has been a major urban or rural problem. Even cities that have been located on rivers have been destroyed by long droughts (usually over ten years). Fresh water has become an intractable problem in the past 50 years, as more aquifers have been drained and more water sources are used for industrial purposes, such as cooling or washing away wastes. Areas of the Americas and Asia still rely on irrigation, and people steadfastly ignore the warning signs of drought and collapse. These long droughts seem to be caused by changes in circulation patterns on a global scale. Many of these changes are the result of variations in the planet's orbit or the solar output. It is unlikely that these can be controlled by human efforts. But we can be prepared for changes, create urban designs to anticipate them, and be flexible and resilient enough to weather them.

If weather change results in rising sea levels, then too much water might be an insurmountable problem and stress civilization to a breaking point.

Experiment: Combating Drought

Can we combat drought? It is everywhere. We could plant trees and shrubs, or other native vegetation to help, but that could take from years to hundreds of years. Water is an immediate problem. We could divert water, but quantity is a problem. We could desalinize ocean water. There have been test facilities and proposals to do this, especially in Africa and the Middle East.

We could create water in labs on a large scale, but it would be expensive. Or, we could recover it from the planet's mantle. Graham Pearson and John McNeill discovered microscopic water droplets in the mineral ringwoodite deep in the Mantle. The total quantity might be equivalent to what is in the Ocean now. Recovering water from there might be expensive, although not as expensive as moving large asteroids.

Or, we could add new water. Where is the unused water? On asteroids or planetoids in space. One scientist recommended moving a moon of Saturn, Enceladus, to orbit the earth. This moon has a huge lake of saltwater under an ice mantle; it spews plumes from the South Pole. Many geysers break through the surface. A moon of Jupiter, Europa, also has ice and water. It might be too large to move, however. If Enceladus were moved and parked opposite the moon, it might be more habitable than the moon (it might also greatly affect tides on earth). Enceladus, like the Earth, Mars, Europa, and Titan, has the highest habitability possibilities after the earth and a high potential for life.

Could we add too much water to the ocean? That is possible, but the dynamic balance of the continents might just let them adjust slowly, with deeper oceans and reduced land area.

It would be smarter to use common sense and monitor water use carefully. Water use could be reduced quite a bit, especially for agriculture in dry areas. We could recover water and clean it. We could plant billions of trees, shrubs and flowers to store and cycle water. Planting would reduce the likelihood of frequent, long droughts, since more water would be cycling in regional systems.

Figure 69. Getting water from rocks.

Experiment: *Strict Planetary Limits or Expandable Thresholds*

Five years ago, I read an article on planetary thresholds (Johan Rockstrom, et al. 2009) and was pleased that science was finally catching up to intuition, mathematics, wisdom, and fear. In the 1970s, many people had recognized that there were limits to growth and to the planet. Donella Meadows et al., Buckminster Fuller, Paul Shepard, Arne Naess, Paolo Soleri, John Cobb, and others recommended many more wilderness areas, lower human populations, radical design, and reduced consumption. Many of the numbers were intelligent guesses, but a few were calculated, e.g., an optimum human population based on ecosystem productivities and an optimum 50–75% wilderness based on uncertainties related to minimum viable wild networks (Wittbecker, 1976).

The Rockstrom article listed 9 boundaries: Climate change, nitrogen and phosphorus cycles, biodiversity loss, stratospheric ozone depletion, ocean acidification, freshwater use, change in land use, aerosol loading, and chemical pollution. These identified "planetary life support systems" are deemed essential for human survival. The authors assert that once human activity has passed certain thresholds or tipping points, there is a risk of irreversible change harmful to civilization. They attempted to quantify how far systems have been pushed, and found that 3 had been exceeded.

These words—boundaries, thresholds, limits—are not interchangeable, which is why most articles before 2009 used the word 'limits.' A threshold is the point where the beginning or ending of something is perceived, that is what we guess or measure a limit to be. Limit is the point where something ends, or a boundary; it may be visible or invisible. A boundary is a line that marks a limit; it may not be the same. Limit is the most basic 'thing.' The problem for us is that not all limits are absolute. The speed of light seems to be an absolute limit. Other limits are relative, such as a tideline, which marks the limit of water; it changes depending on the pull of the moon. Some limits are invisible. Of visible ones, some are fuzzy; the limit seems to grow or shrink. And, we can determine other kinds.

One problem with the nine thresholds is that they mix natural and artificial elements with natural and artificial processes, as well as with the consequences. For example, cropland conversion is a process that leads to biodiversity loss through destruction and extinctions. The process is unwise beyond a percentage of land, but the limit is the extinctions of species as nodes in a web.

Climate change (and of course, the climate is always changing) is a description of the recent interglacial warming, as well as sudden, large-scale increase in heating resulting from the addition of carbon molecules to the atmosphere from burning and industrial processes.

Nine is a good start, and it benefitted people to see boundaries so starkly presented, but the system is complex and requires many more limits.

For example, pollution is a process that incorporates more than aerosols; it also includes plastics, heavy metals, chemicals, medicines, and nuclear wastes. Every one of these elements has a limit associated with it. In addition to carbon pollution, we seem to be getting close to a limit of plastic pollution with the breakdown into nurdles in the ocean, which could potentially cause massive deaths and collapse there.

By contrast, natural pollution is a natural process, from other natural processes, such as volcanic eruptions or algal blooms. These are devastating enough and we should avoid placing cities or populations too close to most of these now known places in the planet.

We have to distinguish what limits are truly global, and whose exceeding could cause change to or provoke collapse of the entire system, and what limits may affect human comfort. For example, CO_2 levels over 1% will affect which ecosystems thrive, but not destroy the planet (which has had higher during the Cambrian). To keep the planet the way we want it, we have arbitrarily—or perhaps expeditiously to avoid sacrificing industry—set a limit at 350 ppm. Then, we noted that we have exceeded that. The actual limit may be 270 ppm or a lower number.

Some of the limits are real planetary limits. The forest cover, or rather the extent of complex, many-tiered ecosystems that control everything from the evaporation of water to carbon fixation, may be critical to keeping water in circulation. We have not found or calculated such a limit, but we can observe its affects in the Amazon and other systems converted to biological deserts. This is related to biodiversity, which we suspect is a limit, but we are not close to find the actual limit in terms of number of species, relationships and communities. Others are moral limits. The planet has suffered extinction rates far higher than those now and rebounded (a younger planet, true). However, the extinctions now are anthropogenic—we are causing them by conversion of wild lands, by overhunting and trapping, and by pollution. Our attempts are aimed at saving the charismatic species that we enjoy watching, such as polar bears, wolves, whales, and moose, and less so for fungi and internal parasites.

The planet needs elemental cycles, such as nitrogen and phosphorus, to keep those elements available for the regeneration of life. And, there are real, as yet quantified, limits to what those cycles contain and how those cycles move. As we found in agriculture, available phosphorus is a real minimum limit—it tends to end up in the ocean. Nitrogen is a maximum limit kept low by natural fixation; our use of nitrogen in artificial fertilizers is like feeding the addiction of a plant and its community. Oxygen percentage in the atmosphere is a real limit. James Lovelock contends that if too many species are eliminated, then even bacteria may not be able to keep oxygen above 19% and the entire planet will die.

Change in land use is not a boundary; it is a process. The boundaries have to do with simplification, extinctions, and productivity. Furthermore, some croplands have increased biodiversity because they are alternative forms of agriculture producing a mature ecosystem, unlike the vast mono-crops of wheat, rice or corn. The real boundaries are soil depth, water filtration, and worm loads, for example.

Fresh water use is not a boundary. It is a quantity with practical and moral consequences, which we should not pollute. But, the local and global water cycle is, that is, the quantity of evaporation is a limit.

Specific pollutions should have limits. Endocrine disrupters should not be evident in any ecosystem at more than 10 ppm. Plastics, which are killing billions of zooplankton, fish, birds and small mammals, should have limits set based on guesses about toxicity and physical effects, such as intestine clogging in complex organisms. Toxicity is amplified in the ocean.

The ingredients of aerosols have to be restricted within limits. Because of its extreme importance in protecting the atmosphere, ozone is an important limit that must be adhered to.

We must keep exploring and expanding our list of limits, as well as important cycles and processes, and make sure they are not exceeded or disrupted by the effects of the activities of civilization.

Experiment: *Are We in a Dying Planet?*

A friend of mine told me that the planet was dying. I replied flippantly that it probably has been dying since middle age (4.5 billion years), but it has 5 billion more years to go. Obviously, however, to both of us the planet is sick in the medical meaning of that word: The new large extinction spasm, caused by human conversion and simplification of ecosystems, along with mass killings and massive pollutions, has caused a serious fever. Although the planet will probably survive, the cool, luxurious planet as we know it will not continue.

Polar bears will go next, if they cannot form hybrids with another bear species. Then lions—most all the species we admire will die out. The planet will be hotter and dryer for millions of years. Maybe new ecosystems will be generated, with a few charismatic species.

But, these words like those of my friend, are meaningless to most people. They may imagine an economic downturn, but not economic self-destruction from the ruination of entire systems and regions. They may imagine having to go back to farming, but not the drought-collapse of cities and early deaths of billions who cannot get or grow food.

How then can we get them to see, and feel, what is happening? This is what has driven me and many of my friends and colleagues to madness. Perhaps a simple analogy (and my apologies for the simplicity) will help. Imagine you are a bacterium in the body of a woman (call her Eve for reference) hiking in a park. You and your fellow bacteria have burdened the immune system of her body. At first things are fine; you multiply and thrive, enjoy splitting and feeding. But slowly, things start to change. The body fights back with a fever and some of you die. But, you keep multiplying and overwhelm the defenses. The body starts to die. This tipping point is invisible to you. But, you are still doing well. Then Eve dies. Things are fine for a generation or so, because new nutrients are available, from formerly vital organs. Most of you are well, although there are some terrible new bacteria (anaerobic) that seem to be taking over your field of dreams. Things get crazy as her body swells and splits open, allowing terrible, large competitors, including bacteria-eating worms and insects, to fight for the remaining nutrients. Then, everything seems to dry up, and most of you die. Some escape into the soil. The wonderful, stable environment has abandoned you. The planet has only turned on its axis five times during this drama.

Now, let's say that we are living in a living planet, in her atmosphere, ocean, and geoforms (call her Gaea for reference). Then our human population starts to explode exponentially, accelerating our conversion and use of ecosystems and resources, including many other species. Carbon dioxide pollution starts raising the temperature of the atmosphere and ocean. At first things are fine …

Is it a fair analogy? The scale is so different. The living biomass of the planet (our Gaea) is over 9 trillion times that of a single organism (our Eve). The planet has 2.1 novemdecillion individuals (with 60 zeroes), including bacteria, whereas the body has only 475 trillion cells, including bacteria; and, the rate of change is not equivalent between an organism and a planet.

But, for the sake of argument, there may be a tipping point after 11,000 years of human modifications, however innocent they seemed at the invention of fire, agriculture, urbanization, or industrialization. We may be past the tipping point, since it may be invisible to us. After all, we have no experience modifying or controlling the entire biosphere. We have noticed a change now, but we pretend that things are still fine and that economics and politics can control wild systems. How soon will we become wise and how wise will we be?

Figure 73. This is only 2 day's trash. Aprilci, Bulgaria.

Part 5: Human Culture

Experiment: *Are Humans Special? Really Special?*
Once it was so obvious: We humans had a special place in the universe,
below the angels. Then as we learned more and more, our knowledge
seemed to dethrone us. There were other stars in the heavens. Then
the earth rotated around a smallish, average star. Then the Copernican
revolution transformed the universe from geocentric to sun-centered and
then centerless. Then our star was found to reside in the distant suburbs of
the galaxy. Then someone found a hundred million new galaxies. And, so
many new planets, some similar to earth. Carl Sagan once calculated that
even if life were rare and only near one sun in a million, life was probably
on 10 million planets. And, intelligent life on over a million in our galaxy.

By fifteenth-century European and Arabic standards, the universe
was a rational order. The human place was prominent, and human life
had meaning and purpose. The biological universe, however, was still the
great chain of being where humanity was a link between the beasts and
the angels. When Darwin linked humanity too closely with the beasts,
cosmology became even less meaningful.

New Research (*Scientific American*, May 2016) indicates that a nebula
creates a star and planets quite often. New solar systems always seem to
have heaviest planets near sun. Our system is an oddity due to history of
entrainment, as the heavy planets snowplowed and rocketed in and back
out (the 'Grand Tack,' plus the 'Grand Attack'). This was a historical development rather than material evolution. The planets did not settle down
in to the current configuration until 3.8 billion years ago. Life began, as
carbon-lace crystals, 3.8 billion years ago. Was this configuration the first
stimulant for life—did it offer some advantage? Life kept going due to the
opportunities and challenges.

The shifting of the gas planets left earth within the life zone distance
from the sun, a zone where the temperature on the surface of the planet
falls between the freezing and boiling points of water. The snowplow effect
left many asteroids and planetoids that contained water, which was released
as they bombarded the planet (water was already inside the planet).

One Mars-size planet hit the young Earth and formed a relatively
large moon that orbits the earth and swings it in a spiral pattern with its
orbit. Because of its relatively large size and closeness, the moon forms the
other half of a double planet with the earth. The moon revolves around
the earth every 27.32 days. The earth and moon both orbit the sun, so the
entire lunar cycle takes 29.53 days. The moon keeps the earth from rolling
on its side, like what happened to Venus. The moon formation provided
energy pulses and stabilized the axial tilt of the earth. The size of the moon

causes significant tides in the ocean (and less so on the continents), and this variation provides energy to living organisms in the ocean, as well as challenges to them for adaptation.

The atmosphere created a greenhouse effect that warmed the atmosphere further. The exhalations of living organisms added more greenhouse gases and increased cloud formation. Clouds increased the reflectivity of the planet, which controlled solar influx.

The moon exerts a gravitational pull on the earth that is stronger on the closer side. This creates a tidal variation in the heights of the oceans; these vary monthly. For many shallow water creatures, amphibians and mammals, it is good to adapt to these tidal variations. Of course, the earth exerts a pull on the moon, also, but it is less dramatic.

Due to its rotation around the sun, and to imperfections in balance, the earth tilts on its axis. This obliquity of the ecliptic creates seasonal variations, to which most animals have adapted. Any changes, even relatively small ones, could be catastrophic for climate (a one-degree change could account for ice ages). Jacques Laskar et al. have documented the importance of the moon on the habitats of the earth. A stable climate needs the influence of the moon; otherwise, there would be immense variations in solar heating of the earth's surface.

Life is also challenged by energy, gravity, and the moon's behavior. That allows it to change. From a cosmological or ecological perspective, living organisms interpenetrate deeply into nonliving forms and the earth. Individual organisms are woven into a complex fabric. Once living beings were formed on earth, their populations must have exploded to use every source of energy available. The fossil record indicates that three billion years ago the earth was inhabited only by blue-green algae. James Lovelock suggests that life optimized the environment for its own use.

Life became more complex. Plants and animals formed life-images of their environments. Some animals, monkeys and wolves for instance, created cultures that were passed down to the young through learning. Human beings expanded their cultures and found ways to put it in writing, as it were. After the creation of agriculture and surplus, the whole complex of civilization—land ownership, technology, engineering, job specialization, permanent buildings, streets, trade, writing, walls—developed. Ideas concerning humanity and the nature of the universe tend to form a coherent system in which ideas are integrated or rejected over days or centuries. Culture includes all of the expectations, understandings, beliefs, and commitments that influence the behavior of human groups.

After all, the displacements and challenges helped to fit us physically and psychologically into a strange, impossibly infinite universe. So, we are special again, after all. By accident and luck, then by work and intent. Maybe we can make a good future if we can learn to live on a wild earth.

Experiment: *Sunshine & Sex: What Sexes are We?*

Once it seemed so obvious, there were two sexes, basically to combine promising characteristics to increase the chances of survival in a challenging environment. Sex enabled individual characteristics to be shared and mixed in offspring. Many simple organisms reproduce asexually, or clonally; they simply reproduce more and more identical copies of themselves, with no mixing of genes and no sex.

Then, life got complex. Flowers have a great diversity in methods of reproduction. Some plants alternate sexes by generation. The sexual form can be male, female or hermaphrodite (sometimes called a 'perfect' flower with both male and female). The Jack-in-the-pulpit plant can undergo sex switching, being a different sex at each life stage. Trees, such as the Alder tree, can also have male and female flowers.

Some fungi, such as mushrooms, can reproduce with thousands of unique sexes, depending on the species.

Then, animals learned. A cuttlefish male can split itself down the middle, appearing to be male on one side and female on the other, simultaneously 'flirting' with another male and a female. A clown fish can turn into the opposite sex. Most snails and worms, and over 20 families of fish (snook, wrasse, grouper, sea bass etc.), are hermaphrodites. Sequential hermaphrodites are most common, but synchronous ones occur. The golden lancehead snake has 3 sexes: male, female and intersex, like a hermaphrodite but with small sex organs on the least dominate side. Female kangaroo has 3 vaginas and two uteri. A clam shrimp can be male as well as 2 kinds of hermaphrodite. The protozoan *Terahymena thermophila* has 7 distinct sexes that can combine to swap genes. A parasite in lobsters seems to have 7 phases of a lifecycle, each a different sex: male, female, dwarf male, clone, and 3 more unidentified. The rule seems to be, the more varieties, the better the odds for mating success.

Humans, alas, are often more complex than plants or 'primitive' animals. They have sex and gender (although here we will overlook the differences in physical and social status). Despite some references to only two sexes in formal Mesopotamian or English literature, most archaic cultures recognized that there were more than two sexes, and acted accordingly, assigning equally-respected social roles to these 'others.'

The world is far more diverse than can be described in simplistic logical dichotomies. Growing knowledge of genetics and DNA reveals that there are at least six biological sexes that can result in fairly normal lifespans (the others result in spontaneous abortion). These biological karyotypes are: The common forms, XX and XY, and the rarer ones, X, XXY, XYY, XXXY, which range from 1 in 500 to 1 in 50,000 births. The significance can be seen in over 7 billion people, where over 3 to 14.5 million people are not strictly male or female.

Possibly your body, brain, reproductive system, family context, social context, and spiritual context could all be different sexes. Also, advances in neuroscience indicate that gender (behaviors shaped by genetics and culture) is innate and immutable about the time a baby leaves the womb.

Many societies have a 2-sex system with archetypal roles. But, many societies recognize other roles, and make places in society for those people.

The Hijras of India and the Muxe in Mexico recognize a third gender. Such are recognized in China, Japan, Nepal, Turkey, Iraq, Australia, Hawaii, Samoa, Albania, Italy, and Afghanistan. In ancient Egypt, the 'sekret' gender was more than just eunuch. Similar categories existed in Dahomey, Maale, and Sakalava people. Many North American tribes, such as the Lakota, Navajo and Chumash, had third and fourth gender roles related to performing work and wearing clothing, making at least four genders: feminine woman, masculine woman, feminine man, and masculine man.

A few tribes had a fifth, related to spirit, for instance, the Chuckchi shamans were 'nonbinary.' The Bugis people of Indonesia divide their society into five separate genders: Two cisgender, two transgender, and "bissu," where all aspects of gender are combined to form a whole (not exactly the same as intersex). The culture supports the belief that all five genders must harmoniously coexist.

The anthropologist B. Malinowski tried to show that almost all of the human culture could be seen as a mechanism to modify and satisfy the sexual needs of an individual. Too narrow. Regardless, the cultural invention of romance allowed sex to become a thing of mystery or beauty, regardless of the number of sexes.

Unfortunately, the more sophisticated we have become technologically, in our global industrial civilization, the less sophisticated we have become socially and culturally. A large number of groups and governments have tried to stuff human sexual diversity into two simple categories, man or woman, penis or vagina, dress or trousers. It is not working.

Everyone! Please grow up. Accept our complexity and diversity, and just go on to the really important stuff, like not blowing up the planet or killing every nonhuman being on it for food or trophies, or because they might bite us. And, instead of replacing the 2-sex system with a 5, 6, 7 or higher number, why not just *not* forcefully classify them. Just accept the diversity and revel in it.

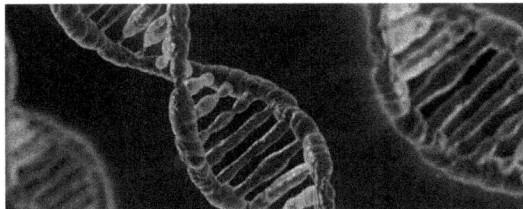

Figure 77. DNA XY

Experiment: *So Which Restroom Should We Use?*
The World Wide Web is actually a good place to find out about human
diversity, as everyone is rushing to display how unique they are to everyone
else. Unfortunately, some people do not care about uniqueness beyond some
important gas-station categories: Restrooms.

In this American culture, for instance, you are either a man or
woman, and 'real' men or women get upset if someone else does not wear
the proper clothes in these rooms (which are usually single person rooms
that can be locked) or even to go into them. Or does not use the proper
position, standing or sitting, once inside (this kind of behavior is sometimes
used in movies to how effeminate the villain is, along with listening to
classical music and trying to discuss important problems like population
pressure, before being defeated with a harpoon through the chest or a bullet
between the legs—yes, I am aware of the contradiction that I watch those
movies, often rooting for the villain, well, sometimes).

On the internet, many people urge the support for and official
doctrine of the two-restroom standard, and spend a lot of clicks damning
those who disobey that 'cultural' rule, making fun of them and trying to
marginalize them with threats of violence.

There are others, suggesting that we try to accommodate other
traditional sexes in the emerging global industrial culture.

Some institutions, from gas stations to hospitals and large businesses,
have attempted to recognize special cases with a third kind of restroom,
sometimes labeled 'Unisex' or 'Family' or 'Handicapped.' Personally, when
I took my mother to the hospital, if it didn't have a third restroom, I had
to take her in to Men or Women, whichever was available, then transfer
her from her wheelchair, help her undress, wait outside the stall, and then
clean her, rearrange her clothing, and transfer her back to the chair. This
was awkward for us the first 20-30 times; it is still awkward for some people
using the restroom before or after us. Fathers and mothers with small
children have similar problems with the 2-category system. As does anyone
taking care of a person of the opposite sex: Which room norm to violate,
mens or womens?

Nobody else seems to have noticed this, but virtually everyone is
covered with clothing during most of the entire process, except babies being
changed. Even during the 20-seconds of exposure at men's urinals, rarely
do men stare at each other's penises. Is there a taboo on brief nonsexual
nudity in shared private restrooms?

The third label, 'Family,' 'Handicapped,' or 'Unisex' does help.
Rather than try to have the 30 or more separate rest rooms, one for each
possible combination, such as 'Woman-but born male' or Male-woman-but-
born woman' or 'Woman helping-male,' which could take up a large part
of an office floor or an airport concourse, why not just start with this third

category, which already exists in many places, 'Unisex'? (I prefer 'Metale' or metasex, since it is beyond 2). Limit its size to one toilet and sink.

We should expect people to recognize that things can be complex, sexually, without the hysteria over differences or remote possibilities. After all, we don't panic when plants and animals do it. Metasex, nonbinary, or unisex? Unisex has already claimed many doors. We could start there, until we can become comfortable with our 5 or 6 (or 7 or more) natural sexes.

Experiment: *Are Nature & Culture Coevolving?*

Nature and Culture are systems. Culture was once called a 'Second Nature,' but human culture has expanded so dramatically that the two systems are better identified as one hybrid wild/developed system, now. The combination of those two systems could be called by a neologism, *domiture*. The fitness of human systems is intimately related to the fitness of species and natural ecosystems. The human attachment to place is critical to understanding why people settled where they have.

Even where the culture may be good, the scale of the culture results in the conversion of wild ecosystems into limited human ecosystems. This is part of a large-scale problem of interference in wild systems. Landscape change intertwines ecological and cultural processes. Environment is the natural; the landscape is cultural. But, it is the same thing. It is not environment or landscape—It is domiture, which includes all agriculture, cities, geospheres, and wilderness.

Stewart Brand thinks that the levels of a healthy society move at different rates. Each operates at its own pace, with the lowest and slowest sustaining the others. Culture moves at the pace of the 'long now,' according to Brand, that is, at the pace of religion. Nature moves in slower evolutionary time.

The study of domiture requires a 'Pan Ecology.' Design has to consider the rates of change of culture and nature, as well as those of emotion and technology. Nature evokes feelings of beauty and terror, joy and sadness—we can participate in wild nature. Nature is fun. It invites play. Ecology can be fun. It invites play. Who cares if nature is not natural anymore? These are only human words and ideas. Nature is healthy, and that is what makes us healthy. Who cares if culture claims dominance? Hurricanes, tsunamis and other events are agents of cultural humility.

Domiture is the entire field of extensions as a unit of study, rather than the reification of something new; it is divided among the studies of natural history, ecology, human ecology, and cultural ecology. Domiture (a neologism made from the Greek word fragments for 'home' and 'again') could do that. The system of culture is embedded in nature; domiture is a larger term to enclose the previous nature-culture dualism. It has to include human reason and human emotion, rather than simply putting them on opposite columns, as with order and chaos, higher and lower, linear and cyclic, as well as agriculture and wilderness. Nature and culture need places to play. Domiture envelops both concepts, giving them room to play. Nature and Culture are systems. Domiture is the combination of those two systems.

How does this happen? Culture does not fit together into a perfectly integrated whole. A culture is a loose-fitting patchwork of ideas, relationships and things. In that sense it is parallel to species adaptation to an environment, but more mutually dynamic. There are discontinuities and contradictions. Humans can tolerate inconsistency and operate with contradictory beliefs: soldiers fight for peace; ministers save the unborn for starvation. If the contradictions become too great and maladaptive, then the culture disappears.

The mode of operation of nature consists of a rhythm of dissolution and reformation. Often the elements of a culture will simply be rearranged by a succeeding culture. A new culture can only be made from the heritage of the old. The International Workers of the World urged its members to make the new world in shell of old one. Our survival depends on our capacity to remake the image of the world from within, phoenix-like.

According to most theories of cultural adaptation (or integration or evolution), resistance to change is normal as a cultural process. Groups like the pygmies have specialized to fit the requirements of the environment, successfully. This makes it difficult to adopt other cultural arrangements that may offer no advantages.

On the other hand, resistance to change itself is an adaptive mechanism. According to Betty Meggers, it works as a successful "cultural isolating mechanism." Isolation remember is what allows a culture to develop in the first place. But, then does it force a culture to become stagnant? Obviously isolation and connection have to be balanced, to allow cultures to remain healthy in a healthy environment. In their domiture.

Nature becomes more complex through filtering, combining, dividing, and mixing. Regular simple dimensions are imposed on nature by culture as necessary fictions. Every form of knowledge becomes a human fiction. Still, that fiction is based on the 'wisdom' of the wild planet, and we are compelled to ask questions: Does nature exist? What is the role of humanity in the destiny of the planet? Who knows? Let's play (wisely of course).

Figure 80. Ready for cattle in Brasil.

Experiment: *Are Humans Too Adaptable?*

Major threats to human life include: Cancer, trauma from accidents, infectious diseases, and parasitic diseases. Also modern diseases of imbalance, such as diabetes, heart disease, and cancer again. Also, polluted water and air, and the stresses from urban living.

Culture adapted to these threats with chemicals, drugs, and emergency treatments, that is, by applying technology. But not with changes in behavior, such as eating less and exercising more, in a healthy environment (not disease-free, but disease-limited).

Humans adapted to many extreme situations, accepting the least satisfactory instead of continuing to strive for a better or optimal set of living conditions. Rene Dubos thought humans were too adaptable. That we could adapt to any level of human impoverishment and misery. To less food and more diseases, to less room and more stress. He seems to have been right.

Cultural adaptation has reached its limits. Now cultures need to change their patterns, local and global. This includes reducing population, based on requirements for healthy ecosystems (with full complements of animals, plants and microorganisms), and restoring or modifying those environments. Sharing a common global identity. Taking personal responsibility for individual health, as well as for the common community health. Participating in the actions taking place in a material world (while ignoring or enjoying the seductions of virtual worlds).

Experiment: *Can Human Diet Force Climate Change?*

For all of our evolutionary past as gatherer-hunters, we ate all kinds of wild animals, from deer and horse, fish and fowl, to dogs and frogs, as many people around the world still do. Some French like frogs' legs and horse steaks, and some Chinese and Koreans like dog cutlets; some Americans eat rat, squirrel and oppossum. During this long evolutionary epoch of human history, our biological nature was shaped by what the regional environment could provide in terms of food, some communities remaining sedentary, others moving with the seasons, migrating with the game. Animal protein was a dietary supplement and not a staple for most early communities except for those in animal-abundant forests, grasslands, the Arctic tundra (where a plant-based diet was impossible), and coastal regions where there was an abundance of seafoods.

While pre-human hominids, like our ape ancestors of today, (with the exception of the chimpanzee), were vegetarians, the first humans were omnivores. They caught and killed their own prey, and stole larger game killed by other predators, whom they learned to intimidate and drive away from their kill. Some domesticated the more tractable game species, like

cattle, sheep, goats, alpaca, and llama, some of which were turned into
beasts of burden, selected for ceremonial and sacrificial purposes, and
for use in draft-work and warfare, like the horse, and the largest forest
animal, the Asian elephant, now almost extinct in the wild, and never fully
domesticated. Domestication of the horse was a major step in facilitating
control and management of free-grazing livestock on grasslands and ranges,
with the assistance of new breeds of herding and flock-guarding dogs.

Coupled with the domestication of plant varieties and land
cultivation, other more sedentary human settlements also domesticated
pigs, cavies, rabbits, wild fowl, and honeybees as sustainable food resources.
Some communities, especially in tropical zones, began aquaculture,
integrating fish production, like tilapia, with a polyculture of plant and
animal production, supplemented by gathering and hunting in surrounding
wild habitats of forest, wetland, bush, prairie and savanna grasslands. These
natural ecosystems they managed with varying degrees of sophistication,
usually with greater care, respect and understanding than those who sought
to colonize, exterminate or assimilate indigenous peoples and exploit their
lands and natural resources with industrial tools.

As human population pressure intensified, along with multiplying and
forever hungry and thirsty flocks and herds, and the need for more land
to cultivate to feed more hungry mouths, conflicts erupted in many parts
of the world. The natural world is being turned into a biological desert or
industrialized wasteland by various human activities. Our singularly most
damaging environmental footprint upon this planet, now recognized and
documented by the FAO of the UN is caused by our collectively costly and
damaging appetite for meat from animals. Some 3.2 billion cattle, sheep
and goats are now being raised for human consumption, along with billions
more pigs and poultry. These extensively and intensively farmed animals
produce less food for us than they consume, and they compete with us for
water. Raising them results in an increasing loss of wildlife and habitat,
and of good farmlands and grazing lands. Linked with deforestation, loss
of wetlands, over-fishing, and ocean pollution, our appetite for meat is the
cause of global warming and loss of biodiversity.

Industrial-scale, capital intensive crop and livestock methods of
production that are heavily reliant on costly and harmful chemicals
and pharmaceuticals. The Amazon Forest and everything in it is being
obliterated by the timber and cattle industries, to grow soybeans to feed
China's pigs and Europe's chickens.

Another example is the dead zone, about the size of Rhode Island, in
the Gulf of Mexico, devoid of marine life primarily because of all the agricultural runoff of animal wastes; from petrochemical pesticides and fertilizers; from fields growing feed for livestock instead of organic food for people;
and from human wastes, much of which should be recycled for farming

purposes if not too contaminated with prescription drugs and concentrated cocktails of contaminating heavy metals, dioxins, and other man-made and man-released environmental toxins.

The production of meat and dairy products, and the wastes and pollution from it, has a great effect on climate. Land conversion releases clouds of carbon dioxide and causes shifts in rainfall and droughts. The animals release more methane clouds, which heat the atmosphere. The industrial production of processed foods, and distribution by ships, planes and trucks, results in poisons and pollution that also heats the atmosphere.

Is it possible to reverse this course? To change our diet to minimize harm to the atmosphere? What would we have to do? First, we would no longer continue to regard meat and other sources of animal protein as a dietary staple because of the enormous costs and harmful consequences of such a diet. A vegetarian diet or a 'conscientious omnivore' diet could reverse the direction of climate change. But that would not be enough. We would have to restore the vast domesticated lands to wild ecosystems. We would have to change the ways we package and distribute food. We would have to use industrial processes that fix carbon rather than release it. Maybe then the climate would be stable and cooler.

Experiment: *Can We Adapt to Surprise with Ecological Design?*
Patterns are not still. A circular pattern through time can be recognized as a spiral (the earth's orbit for example). To be flexible and lasting, a pattern should allow for surprises and discontinuities. The design of ecosystems is vulnerable to surprises because nature is chaotic (unpredictable) and science itself is uncertain (by definition) about patterns of change in ecosystems. Ecological Design tries to anticipate surprise in ecosystems.

The application of Ecological Design can be comprehensive if it includes certain characteristics, from frugality to anticipation. To anticipate means to look forward, to think about possibilities or to expect surprises—in a way to create thought experiments about possible future events. Then possible responses to things can be worked out.

Therefore, design should have certain characteristics pleasing to human perception (and possibly to nonhuman as well), such as interest, movement, and surprise (other characteristics: Change, appropriateness of place, self-ordering, maintaining, edges, and natural cycles). Movement would be basic by rail and highway, but interest and surprise encourage movement by foot and bicycle. Surprise is the result of unexpected vistas or views or arrangements of landscape elements.

A design must also be appropriate. Native species plantings would ensure that exotic animals like hippopotamus would not appear around the corner (although in one sense that would increase the element of surprise). Surprise is unavoidable; it is the unpredictable emergence and novelty that characterizes ecological systems. The Palouse Indians were surprised because they did not have enough information about the Europeans. The degradation of their 'Paradise Creek' was a surprise because the system was complex and we did not predict the decline, or, worse, did not care about it.

Designing, and simply setting aside large areas of habitat, in Reserves, Preserves, Wilderness areas, and Conservation areas, for example, will result in smaller anthropogenic disturbances and fewer ecological surprises.

In industrial societies, people work to modify the environment for maximum use. Working for a maximum requires perfect knowledge and perfect action, otherwise there is disturbance and surprise in the forms of extinction, suppression, soil degradation, water pollution, and collapse. Industrial people are unable to chart the chaotic conditions that arise from the multiplex of variables, from snow and ice, to drought, inversions, cycles, and population variation. But they pretend they can and pursue a maximum management strategy. To avoid too many surprises, we should stop that. Good design can help us adapt to surprises.

Experiment: *Could We Reduce Population Humanely?*

At current levels of luxuries and resource use, the planet is overpopulated. There have been many possible scenarios to reduce the population. In one movie, a villain plans to kill 4 billion to save the earth. A vaccine shot creates a plague that kills 900 million. An escalated war in the Middle East kills 30 million (after 300 million in 100 years). One congressman suggests that we might legally limit couples to one or two children.

Is there a humane way to reduce the population to a target number that would keep all the advantages of size, yet not threaten the stability of the planet? Calculations for a good population have ranged from 450 million to 3 trillion, but for the sake of argument, we can use 1 billion as an optimum population. If not a humane way, is there an effective, inhumane way? Would we choose that if necessary? Regard some of the pop-cultural and academic ways to reduce to 1 billion.

Strategy 1: In one movie, the villain has an electronic device on cell phones to cause people to murder one another. He plans to save just a thousand of the best or richest for rebreeding after everyone literally murders everyone else. A thousand may be a real bottleneck for genetic diversity, so there may be a lot of dangerous inbreeding. It would take a long time to get up to a billion, so cities would mostly decay during that time. Due to logistical problems, however, the population would likely still be over 3 billion before it started racing up to 7 billion again in 28 years. If people survived having 4 billion dead bodies decaying. Outcome: Things might be far worse.

Strategy 2: In one movie a scientist engineers a plague to kill a billion or 2. It has the same limitations as #1, a mess but with 4-5 billion people left to mate and reproduce, the population is back over 7 billion in less than 20 years.

Strategy 3: This is based on the long-term effects (20-100 years) of pollution. We would simply allow weak sperm, as a response to industrial poisons, to lower population to 1 billion in 20-50 years. This would be long enough for society to adjust for burials and urban downsizing, but there is no guarantee that the problem would stop. Outcome: Eventual extinction of humanity from weakness (to industrial toxins).

Strategy 4: Kill people in wars. This would satisfy the 'violent ape' hypothesis by allowing irrational rage. But war is hard and expensive. In the last hundred years, we only killed less than 400 million, at the cost of trillions of dollars and the loss of some great cities and works of art. Outcome: War might kill a billion or 2, then the population might flat line for 20 years, then it would go from 5-6 billion to 7.6 or higher. Nuclear war would be much more effective, but the effects could cause massive extinctions, of humanity and millions of other species.

Strategy 5: Taking a cue from the discredited Chinese experiment, a global government could impose a global 1 couple/1 child limit. But to be effective a large majority would have to agree and be willing to impose sanctions on cheating couples or nations. Laws would also have to control sex ratios, cultural ratios, abortions, and many other factors. India and China have shown that coercive population policies can backfire. Family planning centers would have into fit the mosaic of a culture. Financial incentives could help. The outcome: The population would drop to 1 billion, but immigration and exceptions would have to be granted to balance some nations, which experienced less than the limits.

Strategy 6: Educate women, as well as men and children on biology, reproduction and professional careers. Add incentives for fewer children through taxes, refunds and special attention. License parents. Provide careers for all who want them (with a boost from the massive need for renewing infrastructures, making new cities, and cleaning and restoring domestic and wild ecosystems). Make parenthood romantic, important and desirable. Outcome: World population drops to 750 million to 1 billion. Industrial culture becomes whole and healthy. I would vote for Strategy 6, even if the movie turned out to be boring.

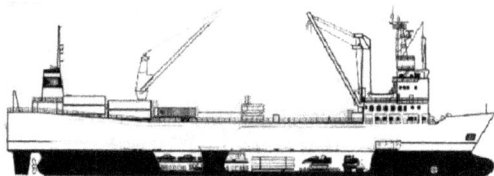

Figure 85. Cargo of death?

Experiment: *Can We Make Laws of Common Sense?*

What are the laws of common sense? They are understandings shared by all reasonable people without the need for debate or compromise. Unfortunately, common sense has a long, complex history, where meanings and applications have been debated in many areas from philosophy and religion to economics and politics. Obviously, superstition, selfish behavior, violence and the irrational are not to be confused with or to be part of common sense.

Some things are obvious. People share a body plan with similar senses and perceptions. People live in communities with a common culture. People have the same basic desires and ambitions to live well and to live better, for themselves and their children.

People in communities bind together to face and accept the unknown. They live together for safety as well as for stimulation. They make standards of behavior and rules of law to specifically make things obvious.

These laws of the community are basic: Do not poop in your own nest, or in other nests close by. Do unto others as you would prefer (the Golden Rule).

Minimize dangerous conditions nearby. Do not keep guns near your bed or where children can reach them. Do not keep rattlesnakes or wolves as pets. Do not act or drive recklessly.

It is common sense to respect your neighbor and not to kill her.

Do not force violence on other people, unless they do it first. Then limit it to brief retaliation (the Tit-for-Tat rule in Game Theory).

Cooperate with as many people as will help you. Others will do this because they benefit.

Understand what your real needs are: Security, stimulation, inspiration, respect, and sources of basic needs from food to wildness.

Although common sense is applied within one's culture, often it is not applied to people of different cultures. It needs to be. People with different clothes and different behaviors must merit the same considerations.

However, we need to extend respect to people we do not understand, or want to know or like, in order for common sense and a few basic legal applications to work on a global level.

Figure 86. Who better
to use common sense?

Experiment: *Would Sacrifice Work?*

The myth or belief that education, discussion and moral conviction will produce over 7 billion caring, responsible citizens who will sacrifice time, money and power to share in solidarity with the masses of poor and sick people displaced from unhealthy and converted ecosystems on distant places and at distant times, is unreasonable and unlikely.

However, on a personal, family or community level, people are willing to give up some things, from 5 minutes of time to their lives, to help others or for the common good. They will sacrifice themselves. For a belief, or an advantage, or a future chance at salvation.

Sacrifice is often understood in the context of war, either military affairs or war understood as the violent clash of two cultures. Citizens are taught that it is noble to sacrifice one's life for peace or for the advance of one's nation. Women are taught to sacrifice for men or families. Men are taught to sacrifice for hardships or war. When sacrifice is argued as a necessity, and perceived as one by most members of a society, the extent of sacrifice is carried by the weakest, poorest and least able. The rich can sacrifice small amounts of money or host important parties.

What would happen if even a minority of affluent people did so? Would some corporations follow to be sustainable? Middle-class and poor often sacrifice their time and health to earn enough money to survive. Could poor people, with their small or nonexistent economic margins, commit to green shopping? Should Native Americans sacrifice their lands for other kinds of wealth? Or for the industrial imperative? Perhaps just a reduction of rich through taxes.

It might be better to encourage sacrifice to remove the roots of discrimination, of inequities, of formal violence, and of war. Such sacrifice might mean more and reduce future sacrifice on the treadmill of economy and war. To sacrifice for health, for restoration of the wild, for peace may be noble, but with so many of us, if we divided it fairly, each personal sacrifice would be small. Would that we could work towards that time.

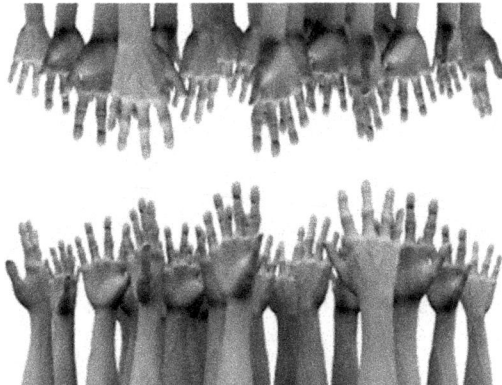

Figure 87.
Voluntary sacrifice.

Experiment: *Are Big Celebrations Important?*

We need new rites of passage, from the personal to international groupings. Perhaps nations could create special ceremonies to celebrate the bringing back of a failing nation. In our nation, we could liberate the land not needed, as suggested by Charles Birch and John Cobb, and John Rodman. We can give that land back to trees and wildlife to restore diversity and richness, and balance humanity and wilderness. And, then celebrate its independence.

We could celebrate restorations. Restoration projects have the potential to save entire ecosystems. Agricultural systems might save domesticated species and a few adapted wild ones. Living ex situ conservation could save wild relatives of domesticates. Suspended ex situ conservation could be highly significant in saving domestic varieties.

The fascination of science, and the successes of science, should be promoted as vividly as entertainment or athletic events—and they are easier for people to participate in, requiring only attention, expression, and a basic education. Some of the best science now is being done in grade schools, to study the surrounding environment. This would recognize our commitment to understanding the basic operations of our planet and the universe.

The ecological design of tools considers tools in every aspect as an extension of the human mind. Tools and things are extensions of human ingenuity. Design examines the effects, short-term and long-term, of the tools, and attempts to balance the trade-offs. Tools have environmental effects, as well as physical, biological, and cultural effects on their inventors and users. Tools have different levels of involvement, as well as levels of intensity. Celebrating tools makes a lot of sense. We need to understand how they work and appreciate them.

Ecological design would work on global and regional scales, as well as the local scale. For example, the Colorado River would be allocated a percentage of water to keep the river and its downstream ecosystems (including shallow ocean canyons) healthy—this may require 50% or more of all the water flow. The remaining water would be divided between resident cultures sharing the river environments upstream. This approach promises a fair way to deal with carbon emissions, toxic wastes, and energy use, also. We could have a celebration every time a river started reaching its delta again.

In these ways we would celebrate our involvement in and respect for the wild ecosystems on which we depend. Giving national parks a birthday party just makes good sense, and it would be fun. Recognizing the birthdays of our political structures and parties could be another meaningful celebration. The planet needs a birthday. It needs a celebration for getting well. We need to begin now! What's not to like and enjoy about a good, timely celebration?

Part 6: Human Migration & Movement

Experiment: *Can Immigration Globally Be Controlled?*
Humans have always moved, to more productive ecosystems and herds, to larger villages or bands, to cities, for opportunities and inspiration, and between nations for security or opportunities. They also moved for excitement, to learn, to rest, or just for fun.

In the past 200 years, migration has been to cities or between nations. And, it has been to escape droughts, discrimination or aggressive, invasive vegetation. Some nations encouraged immigration, to fill their frontiers or the need for skilled workers. Other nations accepted anyone and everyone.

Human populations filled up almost all productive territories, then nations started to restrict immigration, mostly to those who would most benefit the nation. After wars, many people emigrated from their war-torn economies. Some nations accepted many of these people.

Now, almost every nation is wont to accept more than a few from each other nation, because they have to deal with crowding and underemployment. Few nations want a river of political or civil war refugees.

Laws are made, quotas are suggested and enforced, walls are built, and residents protest violently, and often zenophobically. Firm borders are said by many to be necessary. To protect the citizens of a nation. There are limits, and they need to be respected and enforced. Maybe there should be an equation to decide who gets in. Within a calculated number for immigration, such as 30 million, a third for people who have contributed and will give more. A third for political refugees, and a third for random drawings. There might be no social programs for new comers until after 10 years of paying taxes. Then, they could qualify for insurance.

But, what if there were no immigration laws? What if all borders were be open to anyone who could reach them. What if nations pledged a minimum amount of help for all immigrants? What is to fear? Too many people? We have that already. Violent criminals? We have all those already. Would that reduce discrimination? Probably not, but it would be a generous thing to offer.

Figure 89.
Nataf now.

Experiment: *EcoTourism: Threat or Menace?*

Certainly, it is eye-opening to see how people in other cultures live, and living and eating with them can be an adventure. But, the modern packaged tours seem to be exercises in wretched excess, as people look out the windows of their steel tubes between stops at cloned modern hotels and restaurants. A few souvenirs to go with their slides, and these tourists are sophisticated travelers. Our direct experience of the world has become shallow, in spite of faster travel. Travel used to broaden the mind, but now it narrows it. We travel in sealed corridors like boxed goods, comforted by homogenized foods and our own language.

The costs for this kind of tourism are high, especially with the costs of jet fuel, boat fuel, and automobile fuel. Special watered down menus are usually offered, not too spicy, but with volumes of wasted food. Special adventures for the tourists often interfere with wildlife patterns. The hotels create a large footprint on the local ecosystems. In many places, the tourists interfere with traditional ceremonies or ways of life. Tourism has transformed many cultural ceremonies, with the demand for close photographs and souvenirs. Sometimes native peoples play to the tourists and what they expect.

Tourists have romantic ideas about what other cultures should be like. When the reality disappoints them, they fall back on ideal themed vacations, where the cultures can dress up to tourist expectations. The obvious wealth of most tourists has effects on the visited culture, as those people witness tourists eating the best foods while wearing expensive clothing and carrying expensive gear.

We have not evaluated the costs and benefits of all tourism. Should we abolish tourism entirely? Perhaps we could limit the impact of tourists on dry, delicate ecosystems, or other places, by limiting the timing of tourists. Perhaps, after calculating the number of tourists who could visit other cultures and places, we could start a lottery for travelers. This would make the event more valuable to the tourists; maybe they would perceive the people and places as unique and valuable in themselves. Tourism can be a vital part of education and the economy, but its potential for the destruction of ecosystems and cultures has to be limited by good and fair rules.

Figure 90. Pripyat abandoned.

Experiment: *Tourism Limits* or *Leaving Nothing?*

Is Ecotourism possible? While informed Westerners adopt some of the more healthful diets of indigenous peoples, their own governments, and donor, 'philanthropic' agencies, like the UN's World Bank, are working to implant their own industrial agriculture and the Western diet in developing countries to sate the rising demands of the affluent, and the tourist industry, for beef, chicken, cheese, ice cream, and in non-Muslim countries, more pork instead of lentils, chick peas and beans.

Less than 30 years ago, the environment was of little concern to most people. Now it is the primary issue for most people. Calvin Coolidge once said that the "business of America of business." The biggest single business in most of the world, now, is the environment. All farming, most pharmacology, most tourism, and many other "industries" have their basis in the health and beauty of the environment. The environment contributes to the largest share of most gross national products directly or indirectly. but the number of workers and tourists would be limited to the impact on dry, delicate systems.

Forest use zones include culture, recreation-tourism, conservation, fish and wildlife, wilderness, trapping, timber, firewood, and alternative products. None of the uses in these zones should damage the forest.

Already most cultures have been transformed by cash crops, mining, tourists, highways, high-rise housing, and condominiums. Physical disruption has been more extensive than the transition to a UN-directed order could cause. Education could ease a transition.

Education can demonstrate that wildlife is important to people and communities for four main reasons: (1) acsthetic value (beauty), (2) economic value (especially tourist income), as natural capital, (3) scientific value (knowledge), and (4) survival value (health and diversity of ecosystems).

For example, wolves control game animals. This gives them an economic value, because they increase the value of game animals. Wolves have other economic values as well; in some countries, people spend hundreds of thousands of dollars to go and see wolves. Wolves also have an aesthetic value; they are beautiful animals, which is why people like to see them. They are wild and different; this gives them a scientific value. We can study them, to understand the biodiversity and workings of their larger community.

Regarding biodiversity, even St. Thomas Aquinas thought that although one Angel is better than a stone, it does not follow that two Angels are better than one Angel and one stone. As for stones, so for wolves. Finally, wolves have a survival value, simply because they have existed as part of wild places for millions of years. If people cannot learn the need for wildlife conservation, the threatened and endangered species will become extinct. Then, many others will die out. If this happens, human beings will lose much of great value that cannot be replaced.

People in countries without wolf populations might consider taking vacations in areas that had healthy wolf populations. For some kinds of research projects, such as watching in the field and at night blinds, it might be possible to attract students and tourists who would pay for the opportunity to work with wolves (as is currently done is several other countries).

It might be possible for local wildlife totems, representing wolves or elephants for example, to interact with tourist activities. They could introduce and guide small numbers of tourists through the living habitats.

Environmental tourism, however, still interferes with the environment, although there is an effort to educate tourists about ecology. Tourists can crowd or interfere with wildlife, which can become addicted to handouts of food. Is ecotourism a mechanism to provide solutions to the problems of the poor of many countries? Or does it make it worse? Especially if it decreases in volume. Surprisingly, tourist dollars can exceed the value of takings, fish for instance, making tourism income greater than from food or resources for many communities. Tourist dollars can make payments for the protection of biodiversity hotspots. Working tourism can bring income to special projects, as citizen-scientists the tourists can collect data and make observations to contribute to the projects. The numbers of tourists should be limited.

Experiment: *Does Culture Set Traps for the Innocent?*

No culture has developed a perfect balance of human and environmental needs. Some do better than others. But all cultures change and age. As a culture ages, it may become abstract, indifferent, self-centered, and forgetful, suffering rigid rituals and cultural amnesia. Even a culture that fits its adherents and place changes and may become unfit even as the adherents and place also change. Cultures seem to have no limitations of size or kind, although declining mental health may be indicative of some limit exceeded in industrial culture. A culture can grow easily, and temporarily, beyond ecological limits.

Cultures have weaknesses. They hold arbitrary ideas, such as the Algonquian idea of the spontaneous generation of animals, which led to overhunting. They sometimes remain indifferent, as the industrial image of the interchangeable artificial units allows no special value for things and no reason to preserve them. Cultures overexploit their environment.

In many cases it is hard to tell if the destructive use of land preceded ecological problems or followed from efforts to maintain production after an ecological challenge. Cause and effect are hard to separate. The same environment that challenges a culture with some kind of change, also offers opportunities with the change. New resources can stimulate economic activity and increase the level of living.

Cultures develop over time as the groups change and seem to grow larger. Cultures are influenced by climate, resources, and by human ideas about their places and themselves. There may also be larger patterns in human culture. For instance, reproductive success and overshoot of resources seem to occur in the development of a culture. The asymmetry of sex, violence, and ecosystem conversions seem to develop, also.

The growth of a culture can lead into traps. Agriculture is an energy trap, because it allows a higher concentration of energy, that is, higher yields, but then it requires more energy be put into the system to maintain it. Sedentism is a trap. As the population of sedentary communities increased, the wildlife numbers decreased. The productivity and narrowness of food increased, forcing people to rely on agriculture. Thus, there was less possibility of returning to the foraging. People became committed to the new agricultural and urban lifestyle.

Addictions, such as those to foods or oil or money, make it difficult to escape from a trap. Addictions can amplify some emotions, such as fear or hate, especially as they relate to the possible end of the addiction or the threat of that end. Addictions can justify illegal behavior, especially those that seem necessary to continue the addiction. Many cultures are addicted to the illusion of control and power. The US is trapped in the belief that only it, among nations, can bring prosperity and peace to other nations, through gifts, trade or violence.

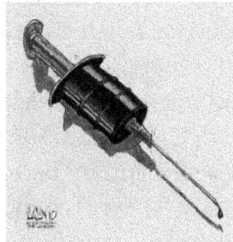

Figure 93. Addiction to Oil
(Credit: Blogspot.com)

How does this happen? Culture does not fit together into a perfectly integrated whole. A culture is a loose-fitting patchwork of ideas, relationships and things. In that sense, it may be a collection of adaptations to an environment. There are discontinuities and contradictions. If the contradictions are too great and maladaptive, the culture disappears.

The mode of operation of nature consists of a rhythm of dissolution and reformation. Often the elements of a culture will simply be rearranged by a succeeding culture. A new culture can only be made from the heritage of the old. The International Workers of the World urged its members to make the new world in shell of old one. Our survival depends on the capacity to remake the image of the world from within, phoenix-like. We could make a global culture that would fit in the constraints and scales of the planet. The planet is wild beyond our imagining, and we need a wild image, a global image that could capture the feel of that wildness.

Experiment: *Living On Comets* or *Viva La Waste!*

Some science fiction writers suggest that a portion of humanity could
live on comets and enjoy whirling around the solar system, with the views
changing for every generation.

Simple calculations by a scientist seem to indicate the superiority of
comets to earth as an address to live. The ratio of total mass M, to available
energy E, or M/E, is R. The bigger R the poorer habitat. According to the
physicist/writer Gregory Benford. But, thinking like a physicist produces
these kinds of errors. Earth R is 12000 tons per watt, while the value for a
comet is 100, much smaller, so much richer, therefore a comet is 120 times
better for life than earth.

Of course, a ray of light would be even better as a place to live.
But light rays and comets don't have ecosystems. They don't have the
minimum of complexity necessary to support complex life forms, such as
ravens, bears, cattle, or humans. Perhaps they already carry viruses or even
bacteria. If so, we could get sick in our artificial systems.

How could we calculate livability in terms of mass? What kinds of life
could a water star support? Bacteria, mammals, humans?

Life needs area, as Benford says, but that requires a minimum area
plus global cycles. Mars, for instance, might be too small for gravity to keep
an oxygen atmosphere. Comets could not host global cycles of elements or
complex ecosystems, or much atmosphere. Both would need sophisticated
artificial systems for human groups to live there long-term.

Energy does not equal life. Mass does not equal life. Life will spread
to where mass is used efficiently, states Benford. Nonsense. There is no
reason to use mass efficiently. There is so much of it. There is no reason to
use energy efficiently. There is so much of that, also. Life is not efficient.
Some waste by some life forms may be food for other life forms, but not
all of it. Waste accumulates at the bottom of the ocean or on the ground
until it is covered up (and eventually transformed geologically). The waste is
staggering, especially with cell growth. Life can only exist where it can waste
profusely. That is not to say that some life forms and some ecosystems are
not efficient, just that efficiency is not the driving force behind them.

Certainly, we will waste food, mass and energy trying to live on a
comet or on Mars (or possibly a water star), but the adventure will likely
be worth it! To avoid waste, we could genetically engineer humans to the
smallest possible size, also for efficiency (although probably not longevity)
in the use of energy and materials (food, matchbox beds, microlighting)—
perhaps mouse-like proportions, although there might be limits to brain size
and operation. We could squeak by on a comet and trade home movies for
a few hours.

Part 7: Human Actions & Feelings

Experiment: *Can We Learn Ecstasy & Otherness?*

Animals present us with related otherness and further human knowledge of ourselves as human beings.

The body gives significance to natural objects, which before they become symbols of concepts, words, or signs, are events that grip the body. The body is in the world as the heart is in the body, but in the heart of the body is the presence of the world, so that the subject is not to be understood as a synthetic activity, but as ecstasy. The body is born with others from the original ecstasy. This ecstasy of experience causes perception to be perception of something (the term is derived from the Greek term, *ekstase*, meaning 'standing out from,' or union).

The world that grasps everyone in its gravity, and which supports their bonds with things and others, is only truly evident in silence and understood implicitly; its alleged positivity, beyond the empirical sense, proves to be ungraspable—the only thing seen in the full sense is the totality from which the sensibles are cut. The foundation of reality is a horizon, from which consciousness carves figures.

The body makes itself a world as things are appropriated into its circuit; but that world is incomplete, as bodily reflection is linked to a standpoint. On seeing other bodies seeing, the limits become more apparent; but these others allow a partial transcendence of the limits through 'intercorporeality.'

It is basic to the Taoist view of process that everything is what it is only in relation to all others. Nature is not a forced order, the result of laws; the laws arising from the motion of nature are conceptual structures projected on nature for understanding and control.

To study that order, a scientific animism needs to understand the meaning of being human, to go below cultural or social explanations of love and alienation. And it may, as a genuine science, forget the analysis and lose itself in the ecstasy of the phenomenon that it sought to explain. Science functions best when we understand so perfectly that we no longer need it.

The greatest human dignity follows from respectfulness of everything as meaningful as ourselves. Such a reverence would treat all substances of the earth as precious, to be used carefully, if at all. It would include all human artifacts, manufactures and societies. It would promote a society where individuals would live in close contact with natural support systems. It would provide each with the aesthetic necessities of life, to develop all capacities. One aspect of an ecological ethic is to live in ecstasy, to help others to see, feel and understand the world and themselves, to feel, touch, smell, savor, and immerse themselves in a wild place.

Describing Empedocles' poetic philosophy, Aristotle hinted that, at that time, philosophic wonder had not yet totally emerged from religious awe, so that cosmology and metaphysics were closer to myth. But Plato allowed for the existence of a passage whereby the mind might cross from philosophy to ecstasy, to achieve insight into the purpose and place of human existence in the universe. What he sought as a philosopher was the meaningfulness of things, which modern science has failed to find so far as an objective feature.

Philosophy, religion, economics, politics, and modern science are lost, because of not thinking in terms of unity. In the presence of nature the feeling intellect of Wordsworth takes over: I feel therefore I think. Wordsworth held that ecstasy is the highest form of thought—I stand out, therefore I am—it is the nearest we get to communication with truth, or perhaps to the idea or feel of heaven. Nature has a secret. Shakespeare knew it: "Ripeness is all."

Experiment: *Attaching To Place—Supraquantum Entanglement*

The Quantum Theory is very successful at predicting events. But, some of it's ideas can be confusing. Entanglement, on the quantum level, is one of them. However, a parallel concept may exist at the scale of advanced organisms. There might be a kind of entanglement. Attaching to place might be Nonquantum or Supraquantum Entanglement.

The planet has many weather systems that intermix across boundaries and scales. This makes forecasting tough. We realize that, similar to quantum entanglement, we have *macroentanglement*. It is too difficult for brains and computers to calculate or untangle those connections. But, if we work and practice, maybe we can think larger and arrange the connections for a good balance of wild wolves and domestic cows, or humans and sharks. Gradually, we might learn to think big.

Raymond Dasmann makes a distinction between ecosystem people versus world people The majority of people are still directly dependent on an ecosystem, and their view of it is different. The Mbuti pygmies view their Congo forest as generous and friendly. The Pueblo Indians see their American desert as a providential home, because of attachment and knowledge. World people are detached and can thrive being mobile and alienated. Ecosystem people live in place.

Every place has a unique identity, a persistent sameness as a result of combinations of factors. Topophilia (Tuan, 1974), love of place, is the recognition that all human beings have affective ties with the material environment.

The attachment to a place is rootedness. Von Uexkull describes the importance of rootedness in his concept of life-world. Simone Weil (1955) regarded rootedness as "perhaps the most important and least recognized need of the human soul." Rootedness arises from participation in a place. It is the need for order, liberty, security, status, and responsibility. A deep relationship with a place is necessary. Without it, existence loses much of its significance. Caring for a place involves concern and responsibility. This attitude is similar to one described by Martin Heidegger (1960) as caring (*Sorge*). Care is the recognition that a human being is a participant in the world. It is tolerance for the essence of a place; absorption in a place; and, concern, the willingness to not change or exploit it.

One ecological benefit of rootedness is that people will take care of a place if they realize they are going to be there for a thousand years. Having a place means that the inhabitant has stock in it and participates in its unfolding, through planting and caring. Detailed understanding of plants in a locale allow gathering of food and medicine. People cultivating a sense of place are people in place. Their work can be appropriate; appropriate growing, logging, mining, or building. People in place acquire a sense of community, nonhuman and human; a shared set of values and concerns; and unexpected health and spiritual benefits.

Adolf Portmann observed that insects and animals displayed a powerful attachment to places—that it was best understood as home. What does it mean to be at-home? The fundamental ambiguity of existence is that humans have different capacities for feelings and awareness. Some feel strongly about a place or home; others never do. Several metaphors have been used to describe the human place on earth. The earth is a storehouse, property, or a spaceship. But the earth is not property, a spaceship or storehouse; it is home. Victor Ferkiss (Fox, 1980b) proclaimed that: "The world and humanity are one entity, one system in equilibrium. Earth is humanity's only home; humanity is one people in relationship to the earth."

In English the term for dwelling is to stay. This is the symbolic opposite of moving or changing. It means to withstand time. Dwelling resists and persists. Permanence is important element in the idea of home. Studies (The Royal Commission on Local Government in England and Wales, in Kaplan, 1983) have shown that people's attachment to a "home area" increased over time. To make good places we merely have to live in them, commit to them and participate and participate in the excitement, and we become fully entangled.

Figure 97. Bats out for the evening.

Experiment: *Can We Kill the Poor?*

Is that a reasonable question? Is the industrial city a jungle? Herbert Spencer saw the world as a jungle, where life was nasty, brutish and short. Spencer coined the phrase 'survival of the fitter.' He based it on the competition in industry, where entrepreneurs fought for money and power. Spencer thought the weak should be eliminated, so he opposed poor laws, charity, sanitation, education, clean water, and pure food, among other things. The less-advantaged disagreed and protested or rioted. Rioting is lashing out at the cage, when people cannot leave cities that do not meet basic needs.

A species or culture that destabilizes its ecosystem through misbehavior risks its own extinction. Human beings make changes to ecosystems, especially urban ones, that endanger themselves.

Industrial society is constantly mobilized for emergencies, in the battles against noneducation, poverty, diseases, and terrorism. Industrial development has never been nonviolent or respectful to people. Industrial production has its own unique selfish style, shape, and scale, and it denies its slaves everything but complacency and despair.

John B. Cobb Jr. notes that realistic hope represents a third alternative to complacency and despair. Earlier, Ben Franklin pithily wrote that those who live on hope alone will die fasting. To hope without directing actions might be too passive an alternative.

Industrial civilization could be retooled to provide for all equally. But, one problem with halting the poisoning of air or the pollution of water or limitation in population is that it would require the most drastic alteration of our present economy. We would have to reduce the gross national product, which is the favorite goal of every administration. We would have to change our styles of living to consume less of everything.

Those in charge, I fear, would rather eat the poor than to sacrifice a pfennig of their own (Jonathan Swift gave voice to something similar when he suggested eating the children of the Irish poor, rather than giving them a minimum of the potatoes that they grew). Eating the poor has never been a formal policy of the "decisioners" (in George W. Bush's verbiage), but I suspect it informs some of their economic cruelties.

I was going to suggest a eutopian alternative of sharing space and food equally, not only in the Middle East, but around the planet. Cobb suggests a thought experiment of his own when he contends that the situation of a typical person would be still unsolvable, if the world was be treated as a ship, instead of a complex living planet. In that popular metaphor, we still would have the problem of people fighting for food, walling off part of the ship, performing triage on the poorest, or perhaps the ship's captain persuading everyone of the necessity of massive action and personal sacrifice. Hope might postpone the inevitable.

Experiment: *Why Does Slavery Still Exist?*
To be enslaved means to be owned as property and divested of freedom and rights, or it means to be completely dominated. The word slave comes from the old Slavic word for the Slavic people, who were among the first slaves to be held in Europe.

Slavery is the economic strategy that people can be held in bondage to perform work. The Greeks thought that their freedom depended on some slavery. Other societies thought that their economies depended on slavery. It was actually rationalized that slavery improved the lives of slaves, since they were often considered to be from subhuman groups. It was also considered to be a personal decision or personal right, although the slave might feel differently.

Different kinds of slavery can be distinguished, even in industrial civilization: Opportunistic, institutional, or comprehensive. Opportunistic slavery was the result of raids or conflicts with other bands in archaic societies. Institutional slavery included labor force collection, as practiced in Egypt in 4500 YBP, or economic slavery for labor collection, as run by the British 300 years ago. Comprehensive slavery refers to the indifferent use of wage earners, animals, and energy, a practice involving a billion people.

Future generations might become enslaved to the decisions of their parents, especially as regards losses and debts. Animals have been enslaved. Machine slaves were the next economic boost. Finally, industrial societies have energy slaves, ten to twenty for each person in some countries. Slavery is based on a number of assumptions, such as contempt for the enslaved, or denial that that slavery is wrong.

Many archaic cultures had slaves, who were usually the victims of conflicts between bands, such as the Kwakwaka'wakw. Slaves became a separate social class in some archaic societies, like China by 5000 YBP. Slaves worked everywhere in Assyrian society, for instance, even running businesses (although forbidden to wear veils). Viking raiders and colonists owned their own land and used Irish slaves for work and for wives specifically (based on genetic evidence). In some African cultures, people entered slavery (and still do as of 2011) voluntarily to pay off a debt.

Internal slavery was a feature of Europe from the Romans to Middle Ages, but disappeared when feudalism disappeared, and labor was no longer the scarce factor in production.

With the 'opening' of the Americas, slaves were considered to be necessary because the acreage increased faster than Europeans could fill it; both were required to develop more wealth. Less than thirteen years after Columbus, in 1505, African slaves were introduced into Haiti. American colonists had tried to force native Americans into labor, but they were still dying from European diseases, which made them unsatisfactory as slaves. So, they looked to another tropical continent, Africa, where the people were resistant to old World diseases. For over 300 years, ten million more people were brought over, in a large trans-Atlantic trading circuit that involved rum, sugar, cloth, and timber as well.

Slavery kept labor artificially cheap. This dampened the incentive to develop new technologies, although new technologies eventually surfaced because they proved to be cheaper than slaves.

The new economics of the industrial age depends on wage slaves, that is, people who need jobs to get necessities in urban areas. After WWII, an overeducated labor pool led to secretaries and then managers, and to the fulfillment of the managerial revolution. Apparently, many human slaves, especially sex slaves, cannot be replaced by animals or machines.

The concept of slavery need not be limited to plants, animals or humans. If we consider our machines as "energy slaves," then each American uses the equivalent of fifty slaves per day. The computer is an information slave and must be the equivalent of at least three other human slaves. Slave owners love the privilege. And the exploitation of inanimate slaves is more easily justified. To some extent, the culture of arts and sciences might not be possible without machine slaves.

But, at some point, slavery corrupts the owners, making them physically or mentally soft. What is the exchange for summoning these information slaves so easily? The failure to develop intellectual ingenuity? Loss of imagination? Lack of trust in intuition? Perhaps writing has changed the function of human memory. Perhaps computers will change the structure of thought and maybe reflective consciousness.

The institutions of slavery raise questions about our relations to each other and to nature. What is our human relation to reality—slave, master, participant, or partner? Cosmology describes the place of a culture in reality. Cultures have often determined inappropriate attitudes towards nature and that resulted in ruin for the air and land. They also defined other people as nonhumans, who could therefore be improved by being allowed to work for true humans. Cosmologies change slowly, although they can be altered faster with sudden changes in consciousness.

Although slavery is officially condemned in virtually every nation, the inappropriate use of people, animals, machines and energy continues, as a necessary part of the agricultural and industrial economies. And, it will continue to be until there is a shift in consciousness that would let us consider the ethical implications of our uses of everything and everyone, especially how these uses perpetuate patterns of domination and inequity, as well as contribute to the degradation of ecosystems and geocycles.

The ideas of dominance and slavery have been integrated into many kinds of designs, from the design of housing and chains to clothing, roads and auditoriums. Changing some of the traditional designs may reduce feelings of inequity and perhaps the actual inequity. The introduction of new stories and myths, concerned with partnership and equality and with the value of other places, species and people, could change our attitudes about them into a healthier context.

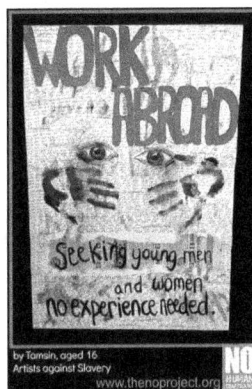

Figure 101. Slavery can start with an offer (Credit: The No Project).

Experiment: *Will An Ideal Help Us to Survive?*

Do our modern ideals help us survive? Our flag, the myths of cowboys, pioneers, astronauts, and market millionaires? Is an ideal or myth maladaptive? If culture is to help us survive long-term, we need to make sure that it addresses the environment. Should we change values, make new laws? Symbolize them with new myths or ceremonies?

We have started to save animals and plants, species and habitats. We have started to design Ecovillages and more efficient automobiles. We have started to create standards for more intelligent buildings and roads. We have started to talk about and sketch whole communities or cities that may fit into specific places.

But, we are not thinking about all cities or all roads, not the planetary patterns that overwhelm places and the vast prehistoric movements of species. We are not paying attention to changes in scale or meaning. We are not considering some kind of economic equalization or the limits to our economic growth.

What kind of mythic or rational ideal should we follow? The cowboy was an effective image for a country that had many frontiers, but the image has become a fashion icon in an urban culture. The pioneer or explorer were also useful when the planet was being explored, and many people can continue to explore, finding new mountains and species, but not many new continents or seas; the ocean, of course is a tremendous unknown and a challenging frontier. The astronaut, for a time measured in only a few decades, was the ideal of courage and scientific and engineering accomplishment. Unfortunately, as discretionary monies were siphoned into the military, the space program was gutted, although several private companies have begun sending up satellites and preparing for sublunar space tourism. Perhaps, the scientist or engineer could become an ideal. Perhaps an outstandingly astute politician.

That ideal, and many others, has been replaced by the market millionaire. Although some people invented and worked their way to fortunes, many others got theirs through one lucky deal or investment. And, many others simply inherited their fortunes, compounded by interest. Some millionaires have led the way for large-scale conservation, like Doug Thompkins, or for philanthropy, like Warren Buffett. Regardless of how aware or bright many millionaires are, they are respected for their money and asked to dispense wisdom and leadership in many areas, arcane or wickedly complex. In fact, their voices are often requested before the explorer, astronaut, or scientist, and sometimes prevail with naïve advice.

So, who is left? Religious leaders, from Gandhi to the Ayatollah, have changed the direction of their nations. Religious Gurus have changed the lives of many people. Political leaders, from F. D. Roosevelt to Barack Obama, have supported programs to increase the health and happiness of their people. But, is this the ideal we should follow? Is there another one that would bring us into sympathy with wild nature and the billions of wild lives? The wise philosopher? A scientifically literate poet? Probably some ideal form will surface in the face of the emergencies and catastrophes that face us in the next 25 years, but the people will choose that ideal, as they always have. We trust they will choose wisely.

Experiment: *Is Humanity a Tragic Species?*
Human beings have been very successful at monopolizing nature for themselves and for their own purposes. This very success, however, may lead to failure, as the tactics that worked so well in an uncrowded world rich in resources are applied to a humanized world with ecological limits. As the Greeks recognized, success can lead to failure, that is, to tragedy.

In Greek tragedy, a single element of value grows cancerously and destroys the whole. The Greek tragedy showed the process by which absolute values swing through reversal to their opposites. The Greek tragic hero was a typical man or woman isolated and projected onto a larger background of fate.

Joseph Meeker judges that the tragic view of life, as embodied in Greek tragedy, is based on a deep conviction that humanity has no part in nature. However, humanity can still be a part of nature and be tragic. Tragedy is more than an ethical conflict. Tragedy can imply conflicts larger than the individual or even society. Ernest Becker claims that the tragedy of evolution is that evolution produced a limited animal with unlimited horizons. In the sense of employing a successful strategy in all circumstances, perhaps natural selection in evolution is tragic—Michael W.

Fox and Garrett Hardin hold this view.

The attempt to avoid enantiodromia, without understanding the operation of nature, ends in tragedy. The definitions of tragedy can be linked to cosmology, the image of the place of humanity in the universe. Tragedy challenges the external order of things, even if the cosmos is only the size of a city. The fatal flaw of the individual is the fatal flaw of the world-view of the individual. This is the root of Garrett Hardin's Tragedy of the Commons—people are locked into a system of self-interest through economic gain without being bound by traditional rules for sharing or cooperating. But, this kind of tragedy results from a failure of cosmology; humans are responsible, not fate or chance. Humans are tragic because they are responsible for their actions. They can choose a tragedy of the commons or of Leviathan—or they can expand their cosmology.

Western images are tragic because death does not imply cultural continuity. Cultural heroes are often mythical beings, of human birth, who restore balance to the people. Unlike theatrical heroes, who precipitate a tragedy by not changing, cultural heroes change to avoid tragedy. For example, many Japanese heroes change their roles as the action develops. The hero, Yamako Takeru for instance, becomes ferocious, then poetic, after success followed by loss, in a way similar to Greek tragic characters. This behavior is called the Nobility of Failure, *mono no arvare*, or the pathos of things.

Despite the arguments of Hardin and Meeker, evolution and ecology cannot be compared to tragedy, which is a description of the fitting of human images and behaviors to a changing environment. In the end, tragedy encourages the acceptance of human limitation. Tragedy proceeds by analogy in the rationalization of the hero. Events are controlled to be consistent with an idea, and end in cumulative catastrophe and purgation.

A complete strategy must include ecosystems and other species, those systems and beings on which our lives depend. We have already made a start in understanding them, as when we suggest that each species has a characteristic face. And, each face is capable of a variety of expressions: Fear, anger, humor, or indifference. Thus, a bear may appear curious, a wolf noble, or a chimpanzee sad. These stereotypic expressions become as masks, although we risk fooling ourselves.

From a cosmological or ecological perspective, there is no tragedy: Everything simply is. Thus, the human situation is tragic to humans only; other species, ecosystems, geological cycles will continue or not. That is not tragic. Although it will be unfortunate for the species and the planet. Unless we are wise enough to avoid it. Responsible human behavior may avoid human tragedy.

Experiment: *Fetus or Physicist*

For many eons, people valued their social group, exclusively. With certain movements, such as the Age of Reason or the Enlightenment, groups of unrelated people became valued equally. Respect and value was extended to the disabled and unconscious. More radical groups extended value to partially formed humans who could not live outside a womb—most doctors agree that a fetus is not viable as an individual until she is about 6 months old (an abortion after that is only to save the life of the mother). For protesters, a younger fetus is considered to have equal value to a physicist, police officer, protester, doctor, teacher, or firefighter.

At the same time, many adult lives were devalued. A doctor (one of many in the past 30 years) was killed for removing a fetus to save the mother's life. Police lives have been threatened for upholding the laws against the destruction of buildings (such as abortion clinics) and violence to the medical staff.

Some groups advocate a no-kill policy in animal shelters dealing with tens of millions of unwanted pets, while other groups humanely destroy excess pets, often to protect songbirds and (beneficial) rodents. At the same time charity groups and social services cannot take care of children in need—and over fifty million children starve to death every year. Some societies take better care of their cars and computers than their neighbors. Other societies cannot provide for a majority of their people.

Why? Is population growth out of control? Do we just care more for the small and helpless (from a genetic predisposition)? Or, have we gone mad? And, our choices are not sensible or rational?

Why would we choose a partially-formed fetus, in which society has a minimum investment, above a 30-year old woman, in which society has invested decades of education and training, and who has contributed years of labor and ideas to benefit society? It seems like a bad trade, based on an uninformed decision. Furthermore, society tries to justify an "either/or" approach on the cheap. Cutting the military just 30 percent (less than the waste reported) would fund all family programs, medical clinics, school lunches, and other programs that support all lives equally. Let's try that!

Figure 104. Physicist at work on stars, 1973. (Credit: UA Lunar Lab)

Experiment: *No Harm?—Swear an Oath!*

Human activities cause a lot of suffering and destruction every year. Millions of mammals are trapped for fur in the US. Millions of unwanted pets are killed every year. Millions of laboratory animals are destroyed after experiments. Billions of chickens are killed for food. A million cattle die from injuries related to transportation. Over half a million animals are shot for sport on wildlife refuges. And, 50 million children die each year from hunger-related diseases.

The activities of forestry cause a large magnitude of suffering: Millions of acres of forest are cleared annually (up to 6 billion trees are killed) and 5 million more are degraded, causing hundreds of species to be extirpated from their homes. Probably each tree is home to 1 squirrel, 5 birds, 30 lichens, 1000 beetles and insects, and 100,000 fungi. Killing each tree kills the habitat for some species and part of the habitat for others.

Most people are ignorant of these facts. Even those that know of them are mostly indifferent. Unlike a Panda or coyote, a tree does not usually inspire strong emotions. Furthermore, habitats are too abstract to attract sympathy. When a forest is considered to harbor snakes and ticks and wet leaves, the reaction may even be negative.

This knowledge, however, is outer knowledge. It does nothing to change our detachment or indifference—Gandhi said that the "hardheartedness" of the well-off was what troubled him the most—or to stop the activities that cause suffering and death. Modern science even argues that seeing emotions in nature is a "pathetic fallacy," although it is more likely that seeing nature dead is an "apathetic fallacy."

The knowledge has to be inner knowledge—the poet Novalis considered that both outer knowledge and inner knowledge is necessary to become a whole person—which can be gained by making yourself part of a place and making that place part of you. This self-knowledge is a modern virtue, although knowledge is rarely considered a virtue. These virtues are similar to the Buddhist eight-fold path to purification and liberation. Other religious figures, such as Jesus and Mohammed, have also preached and practiced following the right path.

Working or living in a forest or by being part of the forest, people can develop virtues such as fortitude and prudence. Virtue is really required for us to work in something that is too complex to know everything about, or possibly even know adequately. Many people are already virtuous, enthusiastic (from the Latin meaning 'possessed with God'), dedicated to solemn acts (from the Latin words meaning to 'vow away'), devoted, and experience awe (meaning 'wonder' or 'fear and reverence,' from the Greek word meaning 'anguish').

These words all seem to be religious terms, but perhaps that is the most appropriate attitude after all. Some kinds of new forestry, such as

Ecoforestry, are concerned with resacralizing forests, with restoring them to their extents and grandeurs, by regrounding science in ethics (ways of living together), and by changing our attitudes from utilization and flat efficiency towards awe and appreciation. Our detachment from trees and other beings has to end. Our participation in the life of the forest has to begin.

Foresters graduating from the Ecoforestry Institute (1993–) are invited to take this oath. An oath is a pledge by which a person swears that she or he is bound, because of beliefs, to perform an act faithfully and truthfully. Like physicians and lawyers (for the forest), ecoforesters are expected to pledge to conduct themselves according to the principles of their profession. This oath combines the 1948 Declaration of Geneva for doctors with the 1993 Ecoforester's Way (coined by Alan R. Drengson).

Ecoforester's Oath

On being admitted as a member to the profession of forestry:

I solemnly pledge to consecrate my life to the forest, as a sacred duty and trust;

The health, diversity, and stability of the forest will be my first consideration;

I will respect the lives of all beings that comprise the forest, and I will continue to learn from them;

I will not harm the forest by taking more from it than it can allow, or more than the vital needs of my community, or by using inappropriate technologies or practices;

I will not permit considerations of economics or politics to interfere with the life of the forest, and I will use only nonviolent resistance to protect it;

I will not use my knowledge contrary to the good of the forest.

I will maintain by all the means in my power the honor of my profession;

My colleagues will be my sisters and brothers, teachers, students, and friends;

I make these promises solemnly, freely, and upon my honor.

Figure 106. Gaia
(Credit: Wm. Washburn)

Part 8: Technology & Art

Experiment: *Could We Generate Optimum Energy?*

Too much energy, raw or in resources and molecules, can ruin ecosystems. But, we have a fantasy of unlimited energy and wealth. Should we strive for a maximum or optimum energy use with coal or nuclear? Or could we produce all our needs with solar/planetary energy?

Our energy needs are surprisingly high, and expensive. The United States consumed a total of 89.456 Quadrillion BTU's over the first 11 months of 2014 (about 29.3 million Gigawatt-hours). To generate this much by solar power would require about 30,000 solar farms (as large as the largest, the Topaz Solar Farm). That would require 6.6 billion solar panels at a cost of $73 trillion. The costs of storage would add to that.

Hydroelectric power would be twice as expensive (possibly $140 trillion), although we would need 20 times fewer dams than solar farms. The lifetime of dams (estimated 20-25 years effectively) is a limiting factor for replacement. And, damming rivers has many large, incidental costs downstream and up.

Coal has so many incalculable costs to landscape health, as well as to human health and the climate, that no new mines or plants should be started; coal should be phased out. But, we use it, despite the real expense.

Would nuclear energy be cheaper? No, it would be over three times as expensive (probably $230 trillion), although each plant would prove 18 times more than each solar farm, so many fewer plants would be necessary. Reactors take a long time to permit, plan and build. Construction is a source of significant carbon dioxide emissions. All of which would compound the basic costs.

Renewable energies, such as solar, have many advantages over large-scale hydroelectric or nuclear plants. First is that they can be implemented on much smaller scales, making them easier to build and place near local needs. This would reduce the cost of transporting the energy (which can approach 50% of that generated). Solar power, in homes or in farms, is very competitive in dry and desert areas (e.g., the SW US, Jordan or Isreal).

These costs all seem high due to the scale in the nation, but they won't go away, and using less expensive means in the short run could cost much more in the long run.

Fortunately, there are ways to reduce them: (1) Cut down on our use of electricity to light cities all night, or to heat or cool or homes 25 degrees from the outside temperature; (2) Stop wasting electricity in homes and factories, since almost half goes through the glass windows.

There might be ways to pay for the improvements. Using under 5% of part of the national defense budget (over $600 million a year). Prices will drop with increases in scale and efficiency. We would try to use longer-life-time plants.

Obviously, to use one source of energy for all needs may not be realistic or desirable (in the short run anyway). We already have functioning plants and dams. Many areas have begun placing solar power, as well as wind, tidal or gravity power.

The current impetus to keep burning coal, oil, gas, garbage, and agricultural 'waste' is not realistic. It is expensive (especially under total accounting) and wasteful. We have the technology for alternate sources of power. We should replace old plants and dams with alternative forms. In cost and efficiency, alternative power is a better bet. And, now is a good time to cut our drastic overconsumption and learn that luxury is not the same as excess, and that happiness has more to do with being than having.

Experiment: *Will We Ever Eliminate Cars?*

There is talk of limiting the number of, and the dominance of cars, as part of carbon emissions, as part of all unusual emissions. Mostly just talk.

We will do anything to keep cars, our private transportation: Power them with solar energy, move parking underground, let certain others borrow them, make them lightweight. But, in the end, they are private devices that require social services to use. Only by being integrated as a pod in a mass transport vehicle might they be saved.

Is the car a form of war? Compared to World War II there are a large number of casualties, there is the slower but nevertheless constant destruction of cities, and there is much pollution, and nature is destroyed regularly in the case of cities being over paved. The casualties from cars number 20 million per year.

Cars are part of the dominant cosmological metaphor: The Universe is a machine. The metaphor of mechanism by Descartes implies that matter is inert and that machines must have a creator, i.e., they are not self-making. Although the metaphor is no longer used in physics, ecology and economics are still struggling with the machine metaphors, which resist change. L. Margulis suggests that is consistent with the major metaphors of the industrial civilization, which are concerned with technology, power, and wealth. We can expect change only when the base metaphor changes.

Cities introduced urban sprawl and suburbs after the world war and after freeways were constructed. No one understood the consequences of private vehicles, or the romance of speed and driving alone, which include the territories of parking spaces and roads, and the constant death and destruction. We need to pay attention to the lessons of alternatives (new trains and buses), size (E. F. Schumacher, *Small Is Beautiful*, 1981) and pace (*The Slow Revolution*).

Experiment: *Will We Ever Deconstruct Roads?*

Wherever people go, they build roads. Unfortunately, roads divide the landscape, often fragmenting whole grasslands or forests. Roads have tremendous effects on wild ecosystems. They divide a system and often isolate species to one side of the road; many amphibians, for example, cannot cross a 3-4-meter width of gravel or asphalt.

Many roads are built only to reach resources for a one-time taking; they are rarely needed again, but their effects remain. Like all roads, even these roads act as beacons for human exploration and impacts and so bring people into contact with almost every ecosystem, including some where our presence is not needed or desired. Other roads are simply in the wrong place and are not needed. Many road parallel others that are deemed too crowded for fast movement. Many of these roads should be retired and deconstructed. Redoing roads would provide many opportunities for work.

It is often easy to make suggestions for roads on a local scale, where people are sometimes intimately familiar with the shape of the land. We need to have local plans for roads that consider the kinds and numbers of vehicles, perhaps emphasizing mass transportation on trains and buses. We need to consider the locations of jobs and homes and entertainments that require different amounts of travel time. We need also to consider transportation and roads on a global level, as major corridors for human movement. And, we need to make coordinated and comprehensive plans for roads in that context. All of these considerations could be considered in new designs.

For example, with most of the population of the planet living in arcologies, the number of roads would be significantly reduced, from 676,750 square kilometers (roughly half of one percent of the total land surface in every country, according to the International Road Transport Union) to 230,000 km^2 (a generous number to be sure, but it could be as little as 50,000 km^2). Over 2 percent of the land is under road surfaces in the United States, with a smaller percentage in Europe, by comparison. Land under agricultural production would also be reduced, from 33 billion km^2 (22 percent of the land area of the planet) to 4.5 billion km^2 (Wittbecker, 1983), partly due to improved practices, partly due to the integration of many kinds of agriculture into the city, and partly due to the carefully limited use of wild populations, without domestication or containment.

What would such a change mean to most people? Probably, few people would have the need for private cars in a walkable city, with mass transportation or other alternatives (e.g., in an arcology). A typical day, for most workers, whether administrating, grading papers, policing, or making steel, would start with a walk to work on tree-lined paths, past local stores and businesses, playgrounds, and microfactories. Work with an interesting, dedicated group. Walking home for food and conversation.

Experiment: *Will Nature Fade Away?*

Will nature fade away now? The new generation, a young participant or consumer, with their peers, will determine the way we treat nature. Generations shift over the years as new knowledge and new styles seem to rule. For a child of 18, having made 648 million heartbeats over 150,000 hours, less than 10% of those has been in any kind of formal education, hardly inspiring for rapid heartbeats anyway; the other 90% has been with family and peers, trying to create an identity and a fortunate path through life (which often works against the education).

One of the new influences is television, which will demand almost 40,000 hours of attention, at least in Japan and the USA. Television will start as white noise and graduate to babysitter and entertainer. To compete with its own past, television creates new kinds of meaning, relying on soap operas, game shows, then comic books, travel, cooking, and finally, for now, sophisticatedly-staged crude reality shows, always searching for the lowest common denominator in the culture. The other cultures of the world, its geography, and natural environments are rarely touched on, unless there is a nude drama in a wilderness island.

Another major influence is the world wide web, which is being filled with the knowledge of hundreds of generations: Courses, dictionaries, encyclopedias, and research. At the same time, social sites allow people to share photographs and lies in depth, as well as to meet up at times and buy and sell goods. Like these first two, on-line games are promoting computer literacy, while ignoring traditional literacy, ecological literacy and image literacy (Wittbecker 1985). Touch screens have eliminated the need for a mouse and keyboard, although consoles are still in use. It has been said that this allows people to find order and meaning in life, and reasons to make ordered lives for themselves. The Web seems to be dedicate towards personal expression and pornography. The knowledge sites, news, or in-depth blogs get little attention (less than 5% of hits, little reading).

Children are trained in many new tools, but not about how to question things or use the tools in the right way and time. They rarely learn to think critically. Formal education above high school has been reduced to training for jobs to get money, ignoring the traditional reason for education: To lead people from ignorance. Instead ignorance is touted as a virtue in reality programs, dating herds and politics. Politicians feel that brave blustering and dedicated decisioning are enough to make a nation great (that and a big-stick military). It has been argued that pop culture makes people smarter, and that crowds (possibly mobs) have a natural wisdom. And, although searching leads to a good horizontal movement, mostly with friends, there is little vertical depth. Thoughts and styles harden. Vertical knowledge leads to maturity, which requires self-knowledge, control and responsibility, which might require discipline, pain or sacrifice. What people look at mostly mirrors their fantasies or fears.

The web becomes a trap, and what a good metaphor that is, for adolescents and children, as well as many adults of all ages. It leads them into a universe of raw data and scattered images. Users learn to pick and poke and move on quickly. The attention span shortens. Nonlinear thinking can offer challenges to the searchers, as can nonhierarchichal strategies and nonsequential flows, but these require discipline to use with critical thought or original extension. The mimicry and mockery on social web sites is intense and personality-altering.

But, television and computer screens do more. They treat wild nature as a source of entertainment, and inform the viewers that they are seeing reality on their screens. They are not! They are seeing a carefully edited and packaged 25 minutes pulled from thousands of hours of looking and researching. Most of nature is slow and peaceful, with only rare instances of violent eruption, usually when the predator catches the prey. The working ecologist can spend hundreds or thousands of hours without seeing her subject, then see it briefly in action (although hidden cameras and distant satellites can track and record animals now). Sometimes having many different screens means that one does not have to go out and learn things in the field. When there are so many choices of rich experiences on the web in seconds or minutes, why go outside for hours or weeks?

We cannot let nature be represented by a set of fading phosphors. Wild nature, in its depth of embeddedness and stretch of differences and dimensions, never fades. There are ways to get people outside. Take them into the real world, for walks or to visit parks. Show them how to join groups, like Artists for Humanity so they can express their voices in motions or paints. Encourage them to join social environmental groups like the Sierra Club, which fights for parks and wilderness. Our bodies and brains were molded by a wild environment; they need that to be whole. If natures fades away, then we will follow.

Figure 111. In a dandelion field. (Altazor Forest)

Experiment: *Identify Goals & Limits in Community Designs*

Our local communities are proud to attract more people and larger industries, but do so thoughtlessly, without regard for the limits of population size or the rate of energy use, without sufficient consideration of the effects on the quality of our lives or on the quality of the environment. Although we make plans for people and their activities, the plans are usually reactions to growth and change. The formal development from planning results in a complex of problems, from pollution to ugliness.

We have always tried to exceed the physical and biological limits of places rather than recognize them and be guided by them. This paper suggests an approach to comprehensive planning based on the biohistory of an ecosystem, the cultural values of the people, and knowledge of the limits for sustainable development. This approach makes the limits explicit and set sustainable goals within those limits. As a synthetic framework, this approach provides for the health of the ecological system, as well as for the health of its human inhabitants.

Most plans address problems, from wastewater treatment to air pollution monitoring. Everything else, from employment to pests, is also considered as a problem, and not a direct effect of the cultural implementation of some technology. Most plans seem to be extremely good at compiling area data, from topographic to climactic. These plans are concerned with determining the adequacy of the infrastructure (utilities, streets, sewers) to support actual and projected population growth. Development plans (water, power) are comprehensive in the sense of seeking to meet all needs of the public, agriculture, and industry. But, they fall prey to all the assumptions of the industrial culture. They tend to be multipurpose with the aim of providing maximum net benefits through management of watersheds, fish and wildlife, and flood control.

A plan should consider the whole system. Communities should be designed for an optimal fit within the limits of the system. Ecological planning considers an optimum population within one ecosystem, although trade for some necessities or luxuries connects it to others. This kind of planning is a conscious adaptation of the benefits of technology to the traditional idea of physical, as well as cultural, limits. Direct observation and traditional knowledge yield far more "information" about the societies of animals than autopsies and mathematical models. An outline of a comprehensive plan is presented, to deal with some of the implications, as well as question them.

1. Identify our place within its natural boundaries. Most places exist in a uniquely identifiable ecosystem, with recognizable boundaries and a unique history and character.

2. Calculate the optimum amount of wilderness to preserve the natural cycles indefinitely. If the current area is less than our calculations, restore the difference and set it aside as a reserve.

3. In the remaining area, zone areas for appropriate use, including conservation, preservation, reservation, and artificial areas (with historical, cultural, and functional importance).
4. Identify the resources needed for human use, including raw materials and the productivity of the areas. This productivity can be used to calculate a base line population.
5. Apply cultural modes—in style, values, and technology—to set limits on technology and population. Preserve the cultural values. Renewable resources will sustain a population longer than energy capital like oil.

As part of the formulation of a plan, we have to examine the natural and cultural histories of a place. We need to understand interactions in the ecosystem, as it was with no humans, as it was lightly settled, and as it is now, dominated by humanity.

The realization of hard realities does not mean a descent into chaos or poverty. We will have more flexibility if we choose our way and salvage much of the industrial revolution, at the same time preserving a good part of our ecosystem and lowering our population and rates of resource use. As technology needs our constant attention and involvement, so does democracy. We need to make the interdependency comprehensive so that we can take responsible action. Many things can distract us: politics, toys temporary rewards. We can change the institutions by altering their legal status.

Civilization can follow a circuitous route, practicing rigorous self-discipline and economy (more than if we had started earlier). There must be fewer people than now. There can be a selective reduction of industries. Large industry might still be needed for certain things (televisions or computers), but the technology must be appropriate for many different cultures. There should be a proper mix of handicraft labor, intermediate technology, and heavy industry. The root problem is how to live with technology in a mature manner.

The goal of planning is community success and personal happiness based on self-reliance in food and shelter, self-sufficiency in agriculture, and self-limitation in size and desires. If human patterns were based on mature ecosystems, civilization would be far more complex; human values would allow for the welfare of humans, animals, plants, and land. We have to be wise enough to be disciplined, to leave wilderness for other beings, and yet to make good places for ourselves.

Experiment: *How Scientists Think* **or** *Incomplete Reasoning*

What is wrong with scientists? They find one decent (or simple) idea and extrapolate it beyond what the scientific framework can hold. For example, recently Fred Pearce wrote in *The New Wild* that we must embrace the 'new wild' of invasive species, rather then be crippled by our old vision of conserving nature that we think emerged through coevolution (prehuman). That is like saying random motion is the salvation of the universe, or that the salvation of nature is competition (wait, that was Charles Darwin's first idea, before he realized that other interactions, like cooperation, were equally important—as less astute biologists and philosophers elevated competition to be the end-all of evolution). Sweeping generalities are made. Many logical fallacies, like name-calling and appeals to authority, are used to support it. Then, strangely, it gets published (like everything these days).

Pearce's good insight is that this process is *still happening*: Invasive plants and animals can initiate new ecosystems on barren land, as well as slowly change disturbed or senile ones. Of course, this has been common wisdom since Henry Cowles (1899), Frederic Clements (1916) and Henry Gleason (1927) studied the role of pioneer species (plus chance and historical development) in the succession of ecosystems. Pearce's big new claim for nature's salvation seems to be that all invasive species can restore all the bombed out places from human occupation. Unfortunately, he is arguing against a straw vision of conservation and restoration that no longer exists, to inflate one natural process into a universal solution to the looming catastrophes of extinctions from human interference with ecosystems.

There is not room here to list all the logical and scientific fallacies in this book. For instance, four of them appear before the book gets warmed up: Considering a fictitious human nature under imaginary circumstances and thinking it is real is the fallacy of "misplaced concreteness" according to A. N. Whitehead. Furthermore, there is a fallacy of the false dilemma, where only two alternatives are claimed and one is considered unacceptable, as in: Either A or B; not A, therefore B. Obviously, there are more than two possibilities. Conservation or invasion are not the only alternatives for ecosystems management; intentional restoration may be a better solution that dumping truckloads of invasive species on eroding landscapes. The popular fallacy of Begging the Question (*petitio principii*) appears throughout. In this fallacy the premises are insufficient to establish the conclusion. And, the Appeal to Authority (*argumentum ad vericundiam*) tries to establish that scientists have all the facts. Not.

After all, succession and restablishment are pushed by alien species arriving to barren or disturbed places, and over time form new ecosystems. Sometimes the alien species prosper in the old system and sometimes they grow into new communities or systems. This is one way life colonizes bare land. Some invasive species are harmless; others cannot take hold, and some disrupt vulnerable systems. Although invasive species can increase diversity temporarily, but unless they can fit or change the system, they fall out of the diversity, which may be worse than before. Not every increase in diversity is healthy or beneficial to an ecosystem.

Invasive species travel to new sites with wind, and water, animal fur, and more regularly with humans. Sometimes they colonize an area, other times they interfere with native species that have been disturbed. Ecosystems can continue to function with invasions, but a healthy native system can repulse the alien plants (R. Daubenmire, 1945). These other times may occur in areas that have been set aside for preservation or are being restored to a healthier state after some disturbance or interference (often human caused). In those cases, such as a restoration of the Palouse Prairie, the invasive species should be taken out so that the native bunchgrasses can be reestablished. Restoration in this case is the attempt to return it to the state before agricultural conversion and biocide decimated the grasslands. Disturbance is always essential to the health of a system, as systems are always developing or collapsing, but constant disturbance can cause collapse if the system does not have a history of it.

The looming extinction of African lions, for instance is an anthropogenic problem, which can be solved with aggressive conservation to keep the entire ecosystem from weakening; we should not wait for American panthers or Indian tigers to invade Africa. We are fooling ourselves to think that our damage can be corrected by adding our invasives to the mix. Nature is too complex for false confidence in simplified solutions from an overworked scientific method of analysis and control, based on a primitive two-valued logic and reductionistic cosmology. Destruction and conservation are wicked problems. We need to be developing a new model of understanding that is comprehensive and nondisciplinary, with the participation of nonscientists and nonspecialists—a holistic approach that treats the current scientific method as a special case for simple problems, much like Albert Einstein's theory allowed Newton's to become a special case that applied to very local environments—but, what would we call this new thing? Panethnics? Metascience? Suggestions are welcome).

Another case of scientific saviorism in another article (F. E. Putz et al., 2012) is buoyed up by similar fallacies: If selective logging, or limited clearcutting leaving some soil and plant structures, then an average forest in the Amazon can retain 76% of its carbon and 85-100% of its biodiversity (or to return quickly)—it could be manageable timber.

Maybe, but only if some old forests are left within striking distance for plants and animals to migrate from. Only if the soil has not been completely eroded, as often happens to bare soil under sun and rain. Only if—I'm sure you get the picture. Furthermore, these will not be old-growth species, since there is no old growth. The raw numbers are encouraging, but the structure and complete functioning takes hundreds of years (for the second emergent layer especially). Sheer numbers count, but it is not a virgin forest, or a complete forest. Praying for, and expecting good regeneration, like praying for rain, is not a good scientific strategy. As an aside, a regenerative forest as it is growing, can pull in over 10 times the carbon and fix it, but that decreases as the forest matures. That temporary boost, however, over a very large scale, could draw down atmospheric carbon dramatically. We should be preserving and reestablishing forests on a global scale, rather than cutting and praying.

In a third case, Arizona State University's Central Arizona-Phoenix Long Term Ecological Research Project, led by Charles redman and Nancy Grimm, concluded a seven-year study of suburban Phoenix, noting that human activities, such as building lakes, canals, pools, and corridors, accidentally created 'brand new' 'novel' 'designer' ecosystems. They suggest that these are resilient systems that can supplement or replace native ecosystems.

They state that cities like Phoenix and Las vegas have changed the city, and its matrix ecosystems, with large amounts of energy and water. While ecologists are studying these 'designer' ecosystems, with groups of plants and animals forming new communities, these are only temporary ecosystems (this is not to say that they should not be studied, just not be expected to thrive for 3 million years). Too much novelty causes stress in ecological systems. As soon as the water from the Colorado River falls, the systems will collapse. Water and electricity are real limits, and cannot sustain these artificial systems for very long.

In summary, human desires to take their favorite plants—from corn and wheat to ornamental flowers and trees, not to mention unwanted invasive passengers—everywhere with them, is an important factor in current catastrophic biodiversity loss—the Holocene Extinction, a Sixth Mass Extinction and the first to be driven by human actions (see *The Sixth Extinction*, Richard Leakey and Roger Lewis, 1995). Human disturbance and transformation occurs on a global scale.

I realize that my complaints and charges could be unfair, that science has a narrow scope most of the time, and it may not be necessary or proper to take a holistic perspective every time. But, now of all times, we need that perspective to stop using every single simple idea as a quick fix, without critically thinking about negative interactions and long-term consequences.

Finally, regardless of these comforting new approaches, we have to act now to restore health, with invasives or without, with inadequate models or without, and with missing pieces or without. Wild nature will continue to operate, but we have the ability to direct and perhaps accelerate some of that process to keep what we think is valuable (as well as necessary for our survival and health).

Part 9: Global Ecological Design

Experiment: *Can We Plan Whole Places?*

In place of a comprehensive plan, this thought experiment creates a deductive, synthetic, conceptual model based on data generated from research on biological productivity, the rates of resource use, and cultural valuation. It also considers minimum wilderness preservation, air and water quality, genetic minima, nonrenewable resources, appropriate technological innovation, the importance of cultural frameworks, adventure, research, beauty, uniqueness, and other intangible experiences.

Classical models work well for simple and isolated natural phenomena, but are not as suitable for complex interactions, especially those with subjective dimensions. Deterministic models work well for things that obey physical laws like gravity, but not when the concepts are extended to ecological systems—systems that are adaptive—or to any system modified by human subjectivity.

Imprecision can be characterized by degrees of fuzziness. Any attempt to classify, categorize, and relate organic groupings results in fuzziness. Abstraction also loses details and is therefore a fuzzy process. Fuzzy sets, however, provide a strict mathematical framework for studying imprecise conceptual phenomena in modeling. Fuzzy sets provide a gradual transition from the rigorous and quantitative to the vague and qualitative. The numbers in this model are fuzzy, that is, they are characterized by a possibility distribution. For example, 'the optimum population is much smaller than 4,254,543' or the 'energy maximum to avoid ecosystem destruction is approximately 1.4×10^8 Kcal per hectare.'

A central planning system is designed to take full advantage of computer applications and find a base unit of measurement. But a computerized information model is only partial; what cannot be quantified, such as feelings or relationships, is often ignored. Computers cannot handle personal observations, sensory impressions, or historical contexts or mythical relationships—just those things used by primary cultures to manage their resources. Although this model quantifies many things that seem nonquantifiable, it is essentially a verbal description.

This model has a small theoretical basis, but it is basically an appeal to action. The model is in harmony with strategies for sustainable ecosystems, the conservation of biological diversity, and aspects of global change. The model attempts to work out plans and policies for long-term environmental stability. The plan describes an architecture of physical and social institutions, that is, buildings as well as politics. The model uses metaphors (such as focus/frame) for stability and planning. It contrasts unconscious growth with conscious planning. The goal of planning is to enhance life—all life, not just human life—so it is not restricted to the human species in the present.

The human population energy of an area is related to land area, productivity, technology, and culture in one algebraic expression (Equation

1). The population (P) is a general number which is calculated by adding the total annual agricultural productivity (in Kcal) to the total annual resources (in Kcal), multiplying that sum by a technological and cultural modifier fraction, and dividing that by the annual per capita requirements (U) for food and resources. The available area (A) is the total land area minus wilderness areas, conservation areas, and other areas to be reserved. The net usable productivity (N) is calculated by subtracting the total unavailable productivity (M) from the net primary productivity (NPP); M includes percentages for below-ground productivity, various wastes, and inedibility. The energy values (E) for resources, water power or zinc, for instance are added. The sum of annual food requirements (Rf) and annual resource requirements per capita (Rr) are combined to make one figure (U). The technological modifier, (T) is based on the use of technology in extending or contracting the food or mineral productivity. The cultural modifier (C) is based on the application of cultural values in determining area and productivity; one example is placing cultural sites off limits to exploitation, even though they may have some valuable resources. Total area, gross productivity, and energy values are known quantities, while the remaining factors must be evaluated in a mathematically fuzzy way.

Equation 1.* $P = \{ [(A \cdot N) + E] \cdot T \cdot C \} / U$

This equation gives us an 'optimal' human population for any mixed area (domestic, artificial, wild, or combined) up to the entire planet. For the total figures for all essentials—food, shelter, clothing, transportation—energy is converted to Kilocalories and placed on an annual budget (which could be averaged over 1, 10, or 100 years). These calculations are used for the purpose of illustration; they are not conclusive or binding.

* The entire equation is expanded in the book *Redesigning the Planet: Global Ecological Design* (Volume 3).

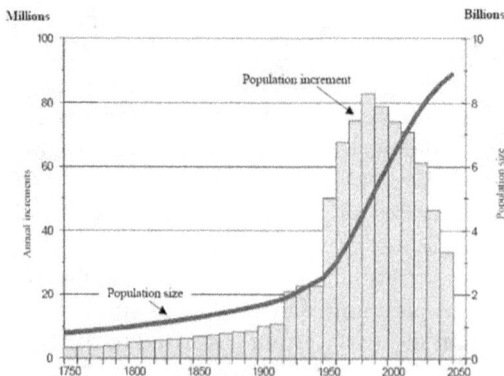

Figure 118. Population growth curve (Credit: Google).

Experiment: *Can We Imagine Better Places?*

Imagination is the act of forming mental images of what is not present or has not been experienced to deal with new experiences. Unfortunately, modern industrial civilization discourages imagination, especially as it applies to large-scale design. The loss of design is the inability to imagine, shape and build things that enhance life and safety; it is the inability to respond to changing circumstances.

There are things we can do to restore imagination, traditional or old-fashioned things like making myths. Myth (or metaphor) condenses a theme, which serves as a symbol for the interpretation of a situation in the world, thus ordering imagination.

Another function of mythology is to provide a cosmology, an image of the universe. In fact, the world has to be recognized and assimilated by the mythopoetic imagination. Myth weaves human knowledge, skills and aspirations together in intersubjective realm of image that blends science and art. Mythic symbols can store and convey vast amounts of information concisely.

Metaphorical expression has an analogical unity in human existence. Its central organizing principle is the single person, with perception and imagination. The expression points the mind toward the broadest meaning. It points, not toward truth, but to an ever-enlarging relational field. It points to the frame, in fact. That which is the frame must be ambiguous. Frames themselves must be used as metaphors.

Karl Popper stated that we must imagine the world to be like a cloud rather than like a clock. The key to the universe is metaphor, not measurement. Clouds are difficult to measure, and even ambiguous to description. But many things can be seen in clouds.

Oddly, we seem to have adequate imagination to describe space colonies and interstellar migration. Yet, when it comes to new cities, human imagination is as limited as human knowledge.

We do not try to imagine the connections between things that we do not know about. Not knowing how trees provide wood or how people cut the trees and process them, we feel no responsibility; we feel no connection. Many organisms exist of which we know nothing. Their worlds have little meaning in a human world. We know what it is to be human, but spend little time imagining other forms of existence. Von Uexkull implies that the human world is only one of the many possible. Although this is true perceptually, the use of symbols allow humans to imagine and represent other worlds.

Maybe the failure of our modern civilization is a failure of imagination compounded with a failure of nerve. We cannot imagine an alternative to war, and we cannot act beyond emotion. Our false models and ideas, combined with our failures of imagination and nerve, for instance, may doom us create only flatscapes and noplaces.

As the adequacy of a habitual view of the earth is questioned, imagination offers simple, alternative ways of seeing. An aesthetic object should not offer a reassuring vision, which interprets or identifies nature, but a naive vision, which surprises, shocks, fascinates or seduces the senses, which awakens desire and stirs the imagination, and which furnishes a feeling of the invisible.

Play is imaginative experience, natural learning entered into freely; education should be more like play than work. Much human activity is play, in place in a community. Vico pronounces that imagination is a turning out of one's self. An imaginative metaphysics shows how humanity becomes all things by not understanding them. The absolute creativity of the imagination diffuses the world from the center of its being, creating an absolute plenitude. Metaphor enlarges the imagination; poetry is the closest approach to objectivity. Through its special position poetry becomes metaphysics, even though embodied, according to Vico. Perception and expression are embodied in the flesh, but imagination is what makes the flesh visible.

Cosmologies adapted humans to limited places, but cosmologies clash and contradict one another. People not only fit physically into a place, by adjusting their eating habits, by making or shedding clothes and houses, they also change their images of themselves and the place so that the two images fit together.

The individual feeds back into the cosmology in altered form what was received. It is almost like a closed loop between cosmology, culture and the individual. Archaic peoples translate the natural world into the language of myth. Being a narrative, a myth is aesthetic as well as intellectual. Myths develop in terms of their own internal logic, drawing together observations of the world. C. Levi-Strauss described the process as bricolage, fitting the bits together, identifying impressions of life as sets and forming them into mythical systems; the world picture is a metaphorical puzzle. Bricolage is the mentality of synthesis, a technique for learning, creating, and expressing understanding, using whatever is available from the past and in the present to achieve an integrating form. This is what mythological thinking does, and what scientific thought might do. So that we can imagine and make good healthy places.

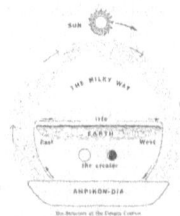

Figure 120. Desana Cosmos
(Credit: G. Reichel-Dolmatoff)

Experiment: *Is Global Ecological Design Possible?*

Global Ecological design is not the additive design of ubiquitous factors such as roads or popular products. Nor is it the design of common structures or symbols, such as cities or money. Although it has to address truly global phenomena, such as the atmosphere or ocean, there is not sufficient knowledge or power to change or reconstruct those structures and processes. So, it has to do mostly with fitting human activities within the framework of the planet. It can minimize the impact of human activities on the global anatomy of the planet, and it can minimize the impact of the planetary processes and events on human constructs.

Global Ecology itself has to be broader that regional ecology, than community ecology, which is broader than human community ecology, which is broader than human ecology. The major difference is scale. Global ecological design is a human process that considers the emergent and unique factors of the planet. We have to change the framework so that we can shape new approaches. We no longer have an external point of view. We are inseparable from the environment and each other, although we can still differentiate and develop new designs.

Global ecological design extends the scales of design from the traditional and ecological to the planetary level, with associated changes in perspectives, for example, in the following sequence: Designing systems of interacting objects within ecosystems of interacting objects, species and nonliving things within a global system (that is noncentered and globally conscious).

Principles of global ecological design reflect the differences between local and global systems. Principles, as fundamental rules, cannot be reduced to other principles, because they emerge from earlier ones and they expand into new forms.

Principles of global ecology, such as stability, contribute to the design of landscapes because those landscapes have derivative principles, or maybe emergent principles. Principles must be flexible to mirror the flexibility of open systems; flexibility is provided by diversity in fact. Many global principles are similar to ecological principles, such as: The Principles of Being, Flow, Separation, Growth, Self-regulation, and the Precautionary Principle. Other principles either emerge at the planetary level or become more important at that level, such as the Principle of Wholeness, which states that a whole emerges from the interactions of parts and that whole is different (not necessarily more) than the total number of parts. These principles result in a further qualification of standards and behaviors, such as 'Protect diversity' or 'Restore degraded cycles.'

All design elements are related psychologically by designers, as focus or frame, as contrast or uniformity, as dominant or recessive, or in a number of other pairs. Good global ecological design means not violating any of the aforementioned principles and ideas.

The pattern should allow for surprises and discontinuities; it can do this if it is flexible. The design of continental forests, for example, is vulnerable to surprises because nature is chaotic (unpredictable) and science itself is uncertain (by definition) about patterns of change in forests.

Global ecological design has to particular attention to emergent global phenomena, such as atmospheric gases (especially methane and carbon dioxide), oceanic currents, and continental drift. It also has to pay attention to global human effects, such as ship, plane and auto transportation lanes, as well as communications technology, from wires and spectra to transmission lines and satellites. And, it has to pay attention to large-scale conversion processes, such as agriculture and urbanization, which change entire landscapes.

Global design can be a bottoms-up process within the limits of global constraints. To use an analogy, architects can build buildings, but the most successful ones are built within the limits of gravity and entropy, as well as within human psychological and social limits. Design has to have the courage or influence to limit production, reduce use, or even reduce human populations and use rates if necessary. It has to do this to integrate the systems in a global pattern.

In order to accomplish a global ecological design, we have to include the spheres of human cultures and global processes. Without a modest redesign of cultures, any actions to balance global designs may fail as cultures focus on immediate local needs, without regard for atmospheric and oceanic cycles, at the least. The totality of human experience and its institutional systems needs to be tweaked. Cultures need to become aware of specific purposes and meanings to become viable as participants in global actions. Cultures have been a form of unconscious self-organization, a shared experiential system communicated between generations. But, they have to incorporate large forms of design, now.

Because global ecological design emerges from and constrains local and regional systems, it is not going to work unless there are hooks at those scales to facilitate those designs, to respond to triggers and traps. If an international organization tries to enforce a top-down decision model for everything, then local and regional decisions will be much more difficult.

We may not want to think of design as the conscious creation of the planet, which is self-organizing, although design may contribute to the creation of a more encompassing image of the planet (literally a world as human-image). Perhaps that image would be of a 'garden,' as Bertrand de Jouvenel suggested. And, certainly human measure is important—it may be the only way we can measure—but, the system is being built on a wild planet, and it supports all human activities with aesthetics and services.

In designing the planet, it is important to design a frame that allows natural processes that are only integrated with artificial processes through larger natural cycles. In this sense we are not concerned with the design of a product or a structure, but a living constraint on natural and artificial systems. In this sense the design is to restore broken linkages, that is, to recreate what was interfered with, to restore the system with native parts or with equivalent niche-makers, and to reduce the interference of human activities.

Globalization as a general trend is triggering a profound shift in human consciousness. Global Ecological Design needs to make hot triggers, that is, the actions have to be immediate: Saving kilowatts, planting grasses or food, or lowering the thermostat. We claim the ability now to geoengineer the planet with large-scale ideas and projects. Even so, there are alternate ways of intentionally reforming the planet: An easy engineering way or a more difficult way of ecological design. If we experiment with engineering ways, we might want to localize the effects, by keeping the change to as few systems as possible, maybe in the southern or northern hemisphere only. For instance, gassing with sulfur dioxide might have drawbacks: Acid rain, and shifting plant, animal and bird life. It might trigger uneven shifts. We rely on technology to 'change the game,' but we can change the game without involving more technology, or more than we have now, by choosing ecological designs that incorporate conservation and frugality.

Ecological design requires understanding of properties and principles that emerge from the local and regional levels; some regional properties cannot be understood without knowing global history. Global ecological design also requires a sixth level of design, which relates everything to the global context; this is the level of global spheres, such as the lithosphere and atmosphere.

Figure 123. Now, Global Ecological Design is needed.

Experiment: *Overcome Wicked Problems with Wild Design?*

In the 1960s, Rittel Horst defined a wicked problem as a kind of social problem which was ill-formulated, with confusing information, too many clients, and conflicting values. Gaia is an example, even more so than social political structures (or human ecosystems), of a wicked problem, due to its history, immensity and complexity. As a system, the planet may be too complex to understand fully, much less manage. Many people have complained that the planet actively threatens human civilization with tectonic changes and extreme climate events.

The living environment is a wicked problem for design, especially now, when anthropogenic changes are contributing to the interglacial global warming. Traditional or ecological design may have difficulty with so large a system. But, global ecological design can address the difficulties at the appropriate level. Global ecological design is as subversive as any art or design; it can overturn some wicked design problems. Global ecological design is a wild way of thinking that can mesh its approach with the wild planet, using an ecological perspective, systems understanding, participation, and standards of knowledge.

We have had great success designing toasters and pencils. We have even had some conscious success designing parks and forests, but for the most part, designs, especially of cities and regions, and of humanity itself, have been aggregated or assembled from all of the small individual and group decisions. They often do not work well together; they do not work well with ecosystems and biogeochemical cycles, either. In fact, sometimes they disrupt international politics or interfere with social systems and cycles.

We have addressed designs on small scales, from the design of ideas and objects to the design of objects in interacting systems and systems of interacting objects. We have not been concerned as much with the design of objects and living beings in ecosystems joined by regional and global cycles. This is the broadest scale that we can design at our current stage of development. We can imagine and discuss redesigning the solar system and the galaxy, but we do not have the knowledge, power or humility to try that, yet.

We need big designs to address at least five major concerns. The first is the planet itself. We need to design land and ocean ecosystems in a way that lets them regenerate without damaging interference. Essentially, we are designing human systems that fit around many systems and interface with others. We need to let the system produce forests and deserts, wetlands and grasslands. Then, we might be able to influence the extent of deserts and forests; we might be able to restore some and create new associations.

Our cities need to be rethought and redesigned secondly. How can we design them so that they do not interfere with ecosystems or watercourses? How can we design them to withstand floods, earthquakes, and volcanic eruptions? If we cannot do that, where should we place them so the threats are minimized? How can new cities withstand three 20-year droughts in 100 years? Old cities could not. Should we have farms on roofs of most buildings? Public vats of food, storage in cans or freezers for 20 years?

The third major concern is agriculture. If people are concentrated in cities, can we simply trade foodstuffs from distant farms or from wild animal harvests? We could reduce our reliance on a few foods, such as corn and wheat that are grown everywhere, often at great expense and great costs to the environment. We could reduce the number and extent of fields, and make them more organic and labor-intense.

Waste and pollution are a fourth concern, now that they are global.

Finally, population growth needs to become planned to become quickly negative to reduce the impacts and interferences on ecosystems.

Wild designs can channel changes through an integrated pattern.

Experiment: *Wild Design Attempts the Global Unknown*

Design should coevolve with wild nature. Local ecological design has to coevolve with its context. Design recognizes these local processes in their regional and global matrix. Wild design has to coevolve with wild nature (xemes) in the planet. Wild design is a primary element to stimulate some possibilities for human creativity.

A framework for design can work at any scale, from a small building, at one end of the scale, to preparing an urban design framework for an entire planet (or possibly solar system) at the other. Models at global scales may be insufficiently realistic. That is why we need local designs, to hang on the design skeleton (a framework). The Potsdam Institute for Climate Impact Research, for instance, is constructing a unified, global-scale model that is tractable and isomorphic. It employs one set of quantitative functions to describe all human impacts, all human adaptations to environmental changes, and all impacts of environmental change on humans—this is where design is important, to limit or modify those impacts with an adaptive kind of design. Within a framework for planning, we need to be able to accommodate the unplanned and the unimaginable. The framework will be incomplete; it will lack detail and definition, but it can be used by all the participants to coordinate actions within physical, ecological and mutual constraints.

Thoughtful rational design works well with many kinds of tools and buildings at a small scale. It may not work with wild systems on a local scale, much less a global scale. We need a wild design. Human agency may be limited, but it could incorporate into its design natural processes that could be effective at the planetary level.

Design works out challenges and problems in an artistic way. Art is wild. We cannot control the effects of art, or even anticipate all of them.

We cannot anticipate the changes it might make. That artistic way is a wild way of thinking and can mesh with large-scale design better than a simple technical approach. Design needs to become wild. Wild design is not human-centered, as most all design in the past has been human-centered. Wild design is based on radical ecology—it is the push, beyond human interests, to consider the character and patterns of ecosystems. Not to subvert or interfere. We can guess what the system 'wants' or how it is developing with reference to its past behavior. We actually know what it wants: To exist, to regenerate. We need to create the conditions for the system to flourish. And, if we use any of it, that use has to be limited to that level of productivity that does not interfere with the survival of the system. We are reintervening in a natural system at different levels, rather than using or interfering for human benefit exclusively.

The word design is modified in this sense to be power with natural processes, not power over them or control of them. Wild design is a conversation across time. We listen, ask, and contribute. We inscribe human stories on the larger stories of the system. Participating means living in the systems. We can reciprocate by giving our bodies back to the system. It cannot hurt also to give our minds to the shape of the system. It is knowing what not to do, as well as what to do, when to do it and where, if we do it. The future is already connected with the past through the present. It just gets complex and unpredictable away from the present. Too complex.

Wild design has to be heroic, especially due to the scale of working on an ecosystem or larger level. Wild ideas are needed for monitoring natural systems, for closing local loops in energy or matter, for working in closely linked webs, and for initiating connections and collaborations. Our cultures, made more intense in cities and by technology like the web, can be the incubators of new forms of wild design.

As the natural balances are upset by human settlements, the settlements also suffer. But, we cannot wait until we learn, or become too large and entrenched to adapt. To achieve an ecological balance, we need to set goals for each kind of ecological spaces. We need additional concepts and goals for the overall system, also, to have the local ones make sense. We also need to have a global coordinating body like the UN to design and implement a global ecological balance. And, at a regional level, nations and communities can go further balancing regions and local systems. Design will have to balance the gains and losses on a local scale. It will have to do it in an equitable manner, with attention to past and current inequities between cultures and peoples, between regions and economies.

Obviously, redesigning local systems means redesigning human structures, paths and influences. Trying to change the patterns of consumption to conserve the capital of nature. Trying to limit human influence. Trying to contain human mass to a limited area of a local system.

Simple conservation makes the most sense, as we just reduce the flows, and have less to bury or deep-place in the ocean, or to avoid with solar screens or aerosols. Efficiency alone promotes more consumption; therefore conservation must limit quantities as it promotes more frugal and fair styles.

We can create thought experiments for all likely situations. We can create possible scenarios for dealing with them. We can create ways to reduce possibilities for disaster or collapse. We have had catastrophes. We may always have them. Bacteria and earthquakes are part of the entire system, as are asteroids and irrationality. And, we have adapted to change. We can adapt to almost anything, Rene Dubos worried, from overcrowded slums to a desert hot world. Design has to define and refine desirable goals and paths. It makes no sense to aim for or to accept a minimum of existence or expression. These approaches will be the best way to reduce the patterns that may influence global warming, extinctions, and ecosystem conversions. And, they may lead to better places for human beings as well.

We humans have never designed complete ecosystems. We have never designed cities, cultures or nations, either. Basing designs on traditional cultures and on the properties and limits of ecosystems should make such large, novel designs useful. Any design also would have to consider political functions at the appropriate levels—personal, community or nation—and maybe regional and global levels as well.

Figure 127. Altazor Wild Design Center

Experiment: *How to Design a Cool Planet*

Since its first molten state, the planet has developed during several stable states, including a hot phase with no ice cover at all and a cold phase with ice caps at both poles. Despite increased solar output and several orbital changes, the cold phase occurred relatively late in planetary history. Humanity as a species developed during the long 'summer' of an interglacial period. Many scientists have noted that anthropogenic changes to the carbon cycle are creating long-term changes to the climate, and that these changes could cause the cycle to skip the next long Ice Age. Several people suggest that a warmer planet would benefit humanity, allowing more northern areas for crops and greater access to resources, especially hydrocarbons; other people fear that the next Ice Age would severely cripple civilization, destroying complexes of large cities and wiping out agricultural lands. Few people would design conditions to cause an Ice Age.

But, which state would we prefer? Which state would be better for the planet? Of course, before now, we never worried that human actions could precipitate one or the other state. Some people argue that we should avoid an Ice Age because of the decrease in available northern land area. Certainly each state can be said to offer some advantages and disadvantages. A warmer planet is thought to have a larger surface area and be more productive, but it would surely have an increase in diseases of plants and animals. A warmer planet could have more droughts; vegetation would create more fibers and toxins, and cold or sub-ice species would die out. Yet, a colder planet that tied up fresh water in ice would have possibly an equal area of land exposed around the equator and in the southern hemisphere. The diversity of habitats would increase. In the past, the challenge of cold has forced changes in some populations; human cultures created more sophisticated tools and clothing in response to cold conditions. Although the productivity of the oceans would likely increase, many life forms in the northern hemisphere would be driven extinct or forced south.

Change has happened, and will continue to happen, regardless of the extent of human impacts and contributions. And, it is unlikely that humanity will be able to exercise conscious control of specific conditions or over the process of change itself. In recent history, design has only had to cope with smaller scale, temporary changes, and not the extremes of a hothouse or Ice Age, although some archaic human groups did expand at the end of the last Ice Age. If the interglacial continues to warm, and the atmosphere becomes more chaotic, a global human civilization will have to alter its food-producing and distributing habits, with extreme ecosystem conversions, and with likely dramatic changes in energy and resource use.

A nonlinear model of the planet has strong positive and negative feedbacks that link the biosphere to atmospheric composition. It is not an equilibrium model. Thus, even if no extra carbon dioxide were produced for decades, the planet would stay in the hot state. James Lovelock says stabilization seems only possible at 5 degrees hotter or 7 degrees cooler than it was about 200 years ago. The transition from negative to positive feedback occurs at sensitive points, not just anywhere. The long-term climate history of the earth shows the existence of several stable (but quite different) states. Present day models do not predict their existence. Climate theory is based on atmospheric physics. Lovelock points out that to be complete, physics has to be based on geophysics, biophysics, and ecosystem science. The earth seems to have two stable states: Greenhouse and icehouse, with metastable states between, like interglacial (there was an unstable extreme 'snowball' earth). The best known hothouse was 55 million years ago, in the Eocene, the dawn of mammals. The recent icehouse ended about 13,000 years ago.

Later, Lovelock concludes that any catastrophe that causes the Gaian regulation system to fail could lead to a hot dead earth, and human actions could precipitate that. Heat-loving plants and bacteria may not have a critical mass of living things to regulate the planetary environment. There is a critical mass of life implied, and that may be related to a critical area inhabited—Lovelock mentions 70-80 percent of the surface, which may not have been maintained during the last Ice Ages, although the low sea level tripled the area of some surfaces, such as Florida in the US. What would happen if carbon dioxide (CO_2) went over 1 percent as a result?

Lovelock also points out that self-regulating systems tend to overshoot a goal and stay on the opposite side of the forcing. If too much heat comes from the sun, the system regulates on the cold side of the optimum. In the past the planet developed this way through a complex web of feedback. For humans to keep the planet cooler, for our comfortable civilization, we will have to manipulate what we perceive as controls or triggers. We might enter another Ice Age, which might be healthier for the planet, but might be equally disruptive to civilization.

Figure 129.
Icehouse planet.

Lovelock suggests that such a colder planet is healthier, more productive and more stable. He refers to the interglacial state as a fever. For life, a cooler earth may be a safer response to solar increase, but there is not a lot of evidence that it was more productive during Ice Ages, even having an equally great land area with vegetation and fewer deserts (or rather a large area of ice instead). Lovelock argues that, because of solar increase, Gaia has greater control during glacial epochs, which has a lower CO_2 concentration in the atmosphere, which he interprets as indicating that the biosphere was healthier and more productive, because cold ocean water is more biologically productive.

Lovelock states this without noting that the oceans are relatively biological deserts, and that life on land may be more critical for cycles. The argument needs to be filled out, since some data of the carbon composition in the deep ocean indicates that there was less organic carbon being fixed. Was the CO_2 too low for plant productivity, even with a larger land surface available near the equator? He notes that the rainforest is an adaptation to recycle water in a warmer environment. And, it is relatively fragile. And now, it is important for carbon sequestration as well. The ocean deposition of CO_2 is important of course, as a physical process in the dissolution of silicate rocks, and as a biological flow towards a sink.

Think about the other consequences: Ice caps cool the atmosphere and lower the sea level. More land is exposed in the equatorial belt, which absorbs more heat. Do trees make the difference, creating more clouds? Are cool ocean currents less cool in glacial conditions? Do they bring up more sediments or less? Is there less sea life than before? Was CO_2 too low for more productivity.

Humanity could be triggering the equilibrium back to a hothouse state. The situation is a catastrophe and requires emergency actions. We need a coordinated massive program of changes. If so, how can humanity avoid a hothouse state and also design and implement the conditions of a limited icehouse state? Reduce carbon dioxide in the atmosphere to be sure. Replant forests. Perhaps even use technology to initiate positive feedback mechanisms that would lead to more ice and faster cooling. Global temperature homeostasis could come about naturally through a negative feedback cycle of phytoplankton, which produces dimethylsulfide, which gives rise to cloud condensation nuclei, which aids cloud formation and thus diminishes incoming solar irradiance and global heating, with an adverse effect on phytoplankton activity. Thus the chain of influences comes full circle, with a net negative feedback effect. The best strategy has to be a combination of conservation, reduction, restoration, and a possibly one appropriate high-technological change.

Smaller, more mobile and adaptive human populations, reflecting a reduced planetary population, would have advantages over large populations trapped in drought conditions or along current shorelines. Anticipating the losses of many cities and fields, populations could be inspired to create new ecological cities, complete with ecological farms and neopoetic mixed ecosystems, in known stable, higher locations.

Experiment: *Could We Redesign the Solar System?*
Science fiction has presented imaginary solar systems in which entire
planets have been reengineered into spheres or bands set an optimum
distance from the sun, or the entire solar system is surrounded by a shell.
Robert Bradbury presented his ambitious ideas in "Under Construction:
Redesigning the Solar System." Bradbury suggests that engineering
feats, such as converting the solar system into a computer (with matter as
software), manufacturing helium stars, or lifting matter from other stars,
may enable Matrioshka Brains or MBrains, comprised of swarm-like,
concentric, orbiting 'computronium' shells that use solar sail-type materials
to funnel and reflect the largest possible quantity of stellar energy. MBrains
could interact with one another, perhaps in a KT-III civilization, which
according to the Kardashev scale measuring technological ability, and
eventually harness the power of and reconfigure the entire galaxy. Bradbury
calculates the computational difference between an MBrain and a present-
day human brain at ten million billion times greater than the difference
between a human and a tiny nematode. With the current billion-fold
difference from that worm, and with our current level of technology, some
less ambitious projects are possible, now.

The first simple thing to be done is to clean up old satellites and
abandoned exploration vehicles. Some of these have already started to
collide with each other and the planet. Some of them pose threats to
new satellites and even to habitations and land. Given the record with
abandoned war machines for the past 120 years, even this seems unlikely.
Cleanup is necessary to avoid accidents and recover useful resources.

We are already aware of the dangers from planetesimals and asteroids
to the earth, especially based on large strikes tens of millions of years ago.
Several governments and agencies have discussed how to monitor their
orbits and prepare to alter or destroy any such threatening small bodies.
We have begun cataloging the orbits of large objects in the solar system,
especially those that orbit the sun and are potentially dangerous, such as
Object 6344 P-L, which was rediscovered in 2007. Modifying the orbits of
meteors, asteroids and planetoids would be wise to protect our civilization
from objects that could dramatically interfere with the stability of our
continents and atmosphere. Asteroids, for instance, could be moved with
spacecraft or lasers—Robert Heinlein and others have written science
fiction stories about this possibility. This level of intervention is possible with
our technologies.

Science fiction writers have imagined moving asteroids and planets
for human benefit. Asteroids like Ceres might be used for water to
transform Mars; Ceres may have the same volume of water as the Earth.
We might also plan to move asteroids and planetoids for minerals and water.

People have dreamed of establishing permanent bases on the Moon and Mars, as well as on other planets such as Venus and Jupiter. One long-term objective of the Chinese Space Program is to create a permanent base on the moon by 2020, for mining and solar energy generation. The United States is also planning a permanent base for 2020, which will serve as a station for reaching Mars. The atmosphere or Mars and the availability of water give that planet the potential to support life originating on earth. Many similarities to our planet, such as day-length and axial tilt, may make terraforming or colonization tempting. Dozens of robotic spacecraft have been launched in the past forty years, although most missions failed. There are economic and political interests in space colonization, besides anthropological ones. Venus has several advantages, including similar surface gravity and shorter launch windows, although the difficulties, such as high temperature and corrosive atmosphere, are much more challenging.

New large constructs might be possible. Ceres, a planetoid between Mars and Jupiter, might be a possible location for a permanent base. It might be possible to construct a new planet between Mars and Jupiter (raw materials should not be a problem). Eventually, assuming we live beyond the age of a typical species of 1-2 million years, humanity may have to move outwards before the sun expands (and then move back when it contracts). In the billions of years before the sun enters a normal expansion phase for a star of its type, we will have time to consider reshaping Mars or building a new planet between Mars and Jupiter—perhaps even moving to Enceladus, orbiting Saturn. But, the logistics and technologies remain quite distant possibilities at the moment. Perhaps at this level our ideas are more extrapolation and guided dreaming. We might also allow concepts of ethics to develop before considering moving entire ecosystems and species onto other bodies or designed artifacts in the solar system.

Some thought has been given to reconstructing the entire solar system, either with new planets or a new system-wide sphere around the sun. When designing a system that large, the first things a designer has to consider are sunlight levels for your area (insolation), gravitational stability, and total power requirement. While it is important to consider the deep future, some of these projects may be beyond the limits of human technology or human psychology.

In a thought experiment, the physicist Freeman Dyson (1959, apparently inspired by a science fiction story by Olaf Stapleton) noted that every human technological civilization has constantly increased its demand for energy. Therefore, if human civilization were to survive long enough, it would demand the total energy output of the sun. Dyson speculated that sufficiently advanced extraterrestrial civilizations would likely follow a similar power consumption pattern and would eventually build their own sphere of collectors. Constructing such a system would make that civilization a Type II Kardashev civilization. He proposed a system of orbiting structures, in a shell or sphere (hence, Dyson Sphere) designed to intercept and collect all energy produced by the sun. Dyson focused only on issues of energy collection, but there were problems with the mechanics of the structure. He responded to criticisms of the concept with qualifications of a "swarm" or "Bubble."

Writer Bob Shaw described a fictional account of the discovery of an inhabited Dyson Sphere in a distant system in his book, "Orbitsville" in 1975. In trying to imagine a more efficient version of a Dyson Sphere, author Larry Niven presented his concept of "Ringworld," a spinning band of approximately the same diameter as Earth's orbit rotating around a star, landscaped on the sun-facing side, with the atmosphere and inhabitants kept in place through centrifugal force and thousand-mile high perimeter wall. When this idea was called dynamically unstable, since the center of rotation could drift away from the sun and eventually let the world come in contact with the sun, Niven used the problem as a plot element in the sequel, *Ringworld Engineers*.

As long as we can eventually be in tune with our own planet, and balance our (soon to be) wonderful cities with restored areas rich in wild associations, we might be mature enough with technology, and its strengths, limits and consequences, to apply it off-planet. That way we can capture the excitement of the solar system and the larger universe, in its awful beauty.

Figure 133. Dyson Sphere (Credit: Danielle Futselaar).

Part 10: Sharing Space—Cities

Experiment: *Why Isn't There a Working Arcology?*

Cities, the largest of human structures, have resisted any conversion to arcologies, to intentional ecological cities. Perhaps the size and investments required for arcologies has discouraged investors and builders from creating them. The start-up costs would be relatively high, possibly $20 billion or more. Energy and agricultural savings, however, should return that amount within ten years, at an optimistic guess.

Perhaps ecology as a complex science is too daunting. Or just the concept that a large, complex building that would incorporate an entire city might seem too much. The concept of an arcology might raise too many questions or issues that have to be resolved before any building could be begun.

Arcological cities might be able to reduce or strengthen connections between people and the environment in a more flexible pattern. Of course, perception is a large part of patterns. We perceive the direction of development as being towards more complexity and more integration until we have megacities in a global society, coordinated on several levels, within a more complex biosphere. An arcological pattern would allow local cities to maintain their diversity and be optimally integrated into new global forms. Self-sufficient cities would allow residents to start thinking locally and acting locally, within the ecological designs of the cities and within local and global constraints.

With an arcology, the city can change its relationship with nature, especially wilderness. Would it determined by the location, shape and size of a wilderness area? Would it be dependent on wilderness? What does it need? What does it take? If it is dependent, how should it guarantee indefinite support? Should wilderness supplement the city and influence every design? Is wilderness the center of life? Is the city? Do cities answer as deep needs as wildness? Could we ever go back from cities to wilderness and villages? People may not want to live in close contact with wild animals and unmanicured landscapes.

The siting of an arcology, with its ability for self-sufficiency, leaves large contiguous areas of wilderness nearby; this would allow people to access wild areas, which would meet their needs for being in nature.

An arcology is a monumental design, which appeals to human desires for creating monuments. And, it could be a work of art in itself, which would satisfy the human aesthetic sense. But, how big could arcological cities be? How big should they be? How big is too big? How small could they be? Is there an optimum size? Is it determined by food production? Cities have been promoted as being efficient concentrations without size limitations, by Stewart Brand and others. But, if cities were organic, like an arcology, then they might have organic limits to overall size and density (as do bee or termite colonies). Aggregation does enhance group survival.

Being optimized for size within a place, an arcology would not be expandable, although that may not be a disadvantage. The outer or inner shape of the city might not be flexible. People may not want to be limited in size or by the necessity to change or customize their city.

Because of the size and complexity of an arcology, new building techniques would have to be employed, on a much larger scale than normal. With new low-carbon techniques and alternative energy collection and generation, the construction of an arcology would not have to rely on traditional technologies, such as concrete manufacture or energy generation from fossil fuels, which contribute to environmental problems, such as carbon dioxide release.

Regardless of the advantages of an arcology, people may be reluctant to move into a new kind of city. It might be the strangeness of living, working and playing in one place. There would be trade-offs of living in an arcology. People would be required to give up many of their habits and addictions, but these would be replaced with new patterns of moving and dwelling that might be much more enjoyable, much more open to a civilized conversation with others.

The shape of the arcology, along with interactions, may not solve the problem of the disparities of wealth. Although the design may encourage interaction and sharing, people will always be able to choose separation. Although the design may reduce crime and violence, people will be able to choose those actions also. Some people may suffer psychological problems from intense living in a planned environment, but perhaps no more so than in any other city. Arcologies may solve some of the problems of increased movements of people, from commuting or emigrating, but not all.

People have expectations about new things. Like any building that relies on complex technologies, regular breakdowns would be expected. How would the arcology allow people to fend for themselves in terms of warmth, energy and food? If the arcology does not have enough internal or nearby external agricultural areas, then food might have to be imported.

An arcology would create a separate ecosystem like an episystem that participates in the cycles of nature, but spins new fabrics and elements, as well as new ideas and adventures. Arcologies complete the idea of the city and make it ecological, sustainable, frugal, and exciting to live in. And, working that out would take experimentation, although people may not want to consciously experiment.

Figure 136. Sketch of the Inground Palouse Arcology
with skylights and water collection systems.
Two small apses of the arcology face south.

Experiment: *How Can an Arcology Form?*

Paolo Soleri notes that, as a human product, architecture has to
be redefined and reconceptualized, to be as antimaterialistic as the
environment itself, and to be created with scientific and religious
collaborations. For Soleri, the city becomes a global term on a planetary
scale. An arcology is a proposal for a habitat for living in uncertain times
and conditions. An arcology is also a construction process for the ecological
city as well as a process for internalizing and transfiguring civilization
itself. A city's vitality as an aesthetic phenomenon depends on abstract and
concrete elements: Miniaturization, equity, harmony, centralization, chance,
complexity, compactness, frugality, the sacred, and integration. In fact, these
properties are similar to those detailed for ecological design.

An arcology is a large dramatic structure that promotes a passionate,
rational way of life. The concept of a one-structure system is not incidental
to the organization of the city, but is central to it, according to Soleri.
Such an urban structure hosts life, work, education, and culture in a dense,
compact system that is interleaved with domestic and wild landscapes. The
compactness of an Arcology leaves 90 percent more land for farming and
conservation.

However, no arcology has been built yet. Apparently, the size of such
a project is intimidating. Funding might be the critical factor, since it may
be beyond the capability of any one investment group or bank. The costs
and benefits have to be laid out over a twenty year time span to show how
effective an arcology can be. The concept has to be sold, also.

The arcology has to be made attractive. It has to bring in people before it is finished, to attract even more people to invest their time in it. It has to find a stable site, not subject to flooding, earthquakes or wind storms. It has to be built where there is land and transport by rail.

It has to show that it can revitalize its location by creating long-term employment. It has to offer a free education at all levels to improve the field for cooperation and competition.

It has to show that it will be efficient, with low energy costs and high density. It would keep density high by emphasizing the elevator, not the car or trucks. Using alternative technologies, perhaps 3% of the area is needed for electricity generation, much of which would be the surfaces of other structures. In a modern city such as Los Angeles, 59% of the ground area is dedicated to streets and parking (other cities such as Detroit, Chicago or Dallas average 44%). By comparison, traditional walking cities were always more compact; the arcology simply eliminates cars, trucks, buses, and trains. However, its walking paths, sidewalks, as well as light rail and escalators are expected to take up 6% of the total area.

It should advertise itself as fun. There would be many recreation opportunities inside. The percentage set aside for parks is higher (at 22%) than for most cities (good cities usually set aside 10–15% of their area). For greenhouse, hydroponic, or small truck gardens, 18% is reserved. The arcology would also be close to outdoor recreational possibilities, as well as within walking distance to wild areas.

Areas set aside for civic and commercial functions are relatively generous. Most of the entire population of the arcology, as well as external workers or visitors would be 'housed' in these areas for at least 6 hours per day. Assuming that each business had 3–9 employees, perhaps 4–11% of the area would be reserved for commercial enterprise. Commercial connections would be many and easy to maintain.

Civic functions would also be relatively dense and so require less space over shorter times. This would be for public meetings, city meetings, or celebrations—but it would also be for heroic meetings or presentations, so the spaces would need to be larger and more dramatic, perhaps 4-10% of the area.

An ecological city like an arcology should be an attractive investment and an option to abandoning cities, whether they are coastline cities vulnerable to climate change, or cities located on geological fault lines, or simply placed on fertile soils. Repairing a massive infrastructure of poorly located or largely undesigned cities is far more expensive than building new cities designed to be in place.

Experiment: *Dharavi Proposal for a Shell Arcology*

With the immigration of rural people into cities, there is a revolutionary change at the global level now that cities house over 50% of humanity. But, traditional cities are failing and cannot handle that influx. Over a billion people live in slums, where they are responsible for their own shelter, food, clothing, and energy. Some slums have over a million people and resemble ramshackle cities—yet, they function!

City Mumbai has one of the densest masses of people on the planet, over 21 million people. The authorized and unauthorized slum colonies (2335 in 1985) of Mumbai contain over 70% of the population. It is a new kind of megalopolis, where most people live in slums, or in huts or tarps around larger buildings and skyscrapers. The slums are vulnerable to fires, floods, cyclones, and many health hazards. Slums are located on hilltops, slopes, low-lying areas, coastal areas, along highways and railways, water mains, and in industrial zones. The city attempts to provide electricity, water and toilets to the authorized colonies (and the unauthorized steal or collect them anyway). Land ownership is problematic.

Dharavi slum used to be the largest in Asia, but now is only fifth largest in Mumbai. It is wedged near the financial district on valuable land. The city wants raze it and move the people north. The people want to stay.

The slum is self-organizing, and has some positive attributes, such as being climate-adapted, very mixed use, and walkable. Social cohesiveness makes some slums livable. Some people have accumulated the aspects of wealth. Many people in slums have color TVs, radios, cell phones, and possibly a washing machine or motorcycle as well. Slums dwellers do not pay taxes or get permits, but some charge others rent. They sell things and services. When slums are in proximity to rich neighborhoods, supply and demand seems to work for maids, drivers, gardeners, and guards. Even in slums, cities allow new forms of income generation. This is not surprising, since cities in ancient Mesopotamia spawned specialists. The density allows cities, even slum cities, to provide more efficient services like education, sanitation and power. More than anything, every slum needs water, sanitation, and electricity; garbage is often recycled internally.

The city has been the most active and dynamic outlet of human life ever developed. Paolo Soleri refers to the city of the future as an arcology to remind us that a well-built city is always ecological architecture. An arcology is a dense city contained in one vertical construction, with a relatively small footprint, embedded in the surrounding ecosystem, and as self-sufficient as possible. For the slum, a new kind of arcology is proposed, a Shell Arcology that would provide a frame for the self-generating order of the slum.

The Mumbai Dharavi Shell Arcology (similar in size to Paolo Soleri's Babel IID, which was designed for 550,000 residents) would be located on a high flat site, with access to roads and railways. It would provide a roof and outer walls, floors, and utilities that would be metered for each floor. Arcologies could offer more security through tenure as well as basic services. The informal economy is very large scale. The basic building would channel people's resourcefulness into building the interior of the arcology.

The arcology would require an internal floor area of 9 million square meters. The footprint would be relatively small at 232,300 m^3 (24 hectares or 58 acres), with 40 stories, 10 of which would be underground in the basalt, 2 within the profile of the ground and 28 above the surrounding grassland hills, which developed in a modified Mediterranean climate. Much of the flat surface would be planted in native grasses; vertical surfaces would have agricultural crops, such as grapes or beans. The density would be 2083 people per hectare. The overall height would be about 85 meters, well within the variations of the topographic elevations.

The Dharavi Arcology would essentially be a beginner's kit to an arcology. Residents would be provided with minimum services and legal security, and even with informal ownership. Utilities would be clustered at points on every floor. Each floor would be completely open; people would be expected to build their own walls to claim spaces. To succeed, the Arcology would have to be presented and supported by central government departments, state officials, municipal authorities, the corporate sector, and finally, mass media.

The Mumbai arcology would be a self-contained habitat of reduced size, with access to the surrounding city and some wild areas. As a single structure, it would integrate the fundamental components of life—dwelling, eating, learning, working—into a frugal form, while accepting a certain amount of ambiguity and disorder. It would be a laboratory for doing and learning, tested through experience, and tempered by the opportunity to fail.

The arcology would create a synergy between habitat and nature, body and spirit, and work and leisure. It would generate the urban effect more intensely, which should energize the inhabitants. The principles of an arcology, from using marginal land and imploding spread to habitat density to internalizing and miniaturizing the urban effect and linking the habitat, energy and environment, would serve as a model for a stable, exciting construct. The creation of a frame for an arcology in a dense slum area would be a good test to see if the self-organization abilities of people, if provided with a basic shell offering basic services, could create a comfortable, attractive, working city.

If the arcology becomes anything like the existing slum in organization, then most areas would be mixed-use, with work areas near living areas. Because cell phones allow such flexibility and connectivity, the infrastructure may not have to be as formal.

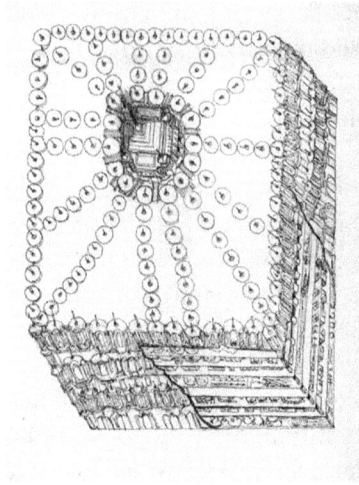

Figure 139. The Mumbai
ShellArcology (Credit:
Wm. Washington).

Experiment: *Invite Wildness In: Wild Mountain Arcology*

The Canadian Wild Mountain Arcology would be built into the foothills
of the Rocky Mountains in eastern British Columbia on the west slope—
Shaped like a ridge itself, it would extend into the plains. From a distance, it
would appear to be part of the mountain, although some regular features,
such as the its peak and curvilinear windows, would be evident. One of the
principles of arcologies is to use marginal lands. Another is to fit into wild
ecosystems when possible.

Designed for 100,000 people, it would require an internal floor area
3 million square meters. The footprint would be relatively small at 40
hectares, with 10 stories, 3–5 of which would be underground. Much of
the lower surface would be planted in native grasses. The density would be
1500–1900 people per hectare (Manhattan is about 200/ha). The height
would 10–30 meters, well within the variations of the mountain slope.

The arcology would offer generous spaces for public activities as
well as external parks and seasonal food growing areas. These might be
contracted or expanded as the population changes. Large wild areas,
especially wild rivers and montane areas, can be reached within an hour of
travel along established roads and rail lines.

The arcology would have unique external features. Except for the
west-facing entrance apse to the arcology, the arcology would resemble a
large ridge. The surface would slope at about 55 degrees. A stream running down the slope would be channeled through the arcology and create
a waterfall for the bottom two stories, before continuing into the local river.
The surface would be rock embedded in concrete. The rock would be of
variable sizes and house hollows for birds, bats and small animals. Many
kinds of lichen would colonize the surface within years. All three sides
would have regularly-spaced nonreflective glass windows for admitting light

and allowing people to observe from their apartments. Some windows could be opened; then, the rest of the rooms could be sealed off from weather or visiting animals. Every group of apartments would share a large balcony, for plants or chairs.

The upper two stories, which would house public meeting areas, would be constructed of fitted rock. These would house the largest open rooms, which would be supported by a kitchen and offices.

The lowest two stories would be open to some animal traffic. The apron of the arcology would have a cover of soil; the depth of the soil would range from 1 foot to 3 feet and be planted with native grasses. The west-facing apse entrance would have a tile face over concrete. The inside of the apse would have a 50% glass cover. The apse transitions to the space of the arcology, starting with the public atrium.

The road leading into the arcology would enter a tunnel a kilometer from the structure. In fact, an elk trail running cross-slope to the arcology would continue under the front part of the building and over the tunnel; it would be most active in early autumn. Other animals, including woodland caribou and brown bears, would be passing visitors. Hawks and Ospreys would nest nearby. The area around the arcology would have coniferous trees (Douglas-fir, pine, larch), native plants such as kinnikinnik, and wildflowers including bunchberry, rhododendron, and ragworts. Wild areas would be within walking distance.

For greenhouse, hydroponic, or small truck gardens, 18 percent of the floor area is dedicated. Individual apartments may also plant herb or limited gardens with fruit or nut trees. A vertical farm would create 'closed loop agricultural' ecosystems, where nearly every aspect of the farming process is recycled and re-used, especially water and nutrients. Using alternative technologies, perhaps 5–15% of the area is needed for electricity generation, much of which would be the surfaces of other structures. A small, self-contained, buried nuclear device could power the entire arcology.

The arcology is not very remote. It is near the city of Golden and close to the Trans-Canada Highway. It would have advanced communications, such as the Internet and cell phones. The arcology would host living, work, education, industry, and culture in a dense, compact system that is interleaved with domestic and wild landscapes. Its shape and density would generate an intense urban effect to energize the inhabitants to be productive and creative.

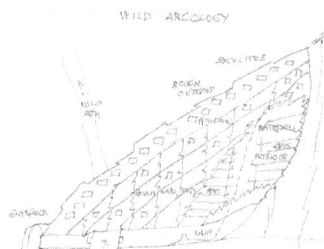

Figure 141. Rough sketch of the side-elevation of the CWM Arcology.

Experiment: *Antarctica World Pyramid Arcology*
Antarctica is a medium-sized continent. It is situated over the South Pole
almost entirely south of the Antarctic Circle. It is a very rough circular
shape with the long arm of the Antarctic Peninsula stretching towards
South America. There are two large indentations, the Ross and Weddell seas
and their ice shelves. The total surface area is about 5.5 million square miles
in summer, approximately twice the size of the Australian continent. In the
winter Antarctica doubles in size due to the sea ice that forms around the
coasts. The average thickness of ice on the Polar Plateau is 7000 feet, and
ice depth can reach 15,000 feet. Ice is an important resource for the planet,
and it accounts for 70% of the fresh water on the planet. Several people
and groups have proposed larger settlements on the continent.

The United Nations has worked to protect the continent from development for its resources. Here, a world arcology is proposed for Antarctica
as a new headquarters for the United Nations. Housed in a single large
pyramid in a neutral territory, the structure would become an iconic symbol
of a new world order, where representatives of humanity could dwell in an
artificial, sustainable, zero-carbon, low-waste ecology within a wild environment. From a distance, as a whole, this city would be a monument with
symbolic value. The arcology, however, would be located in one of the rare
snow-free areas, close to the ocean for relatively easy access by ship.

The arcology structure would be a pyramid with equal bases. It would
be oriented with one edge facing north. The external surface would be a
curtain wall of glass and stone. The façade would be dark-colored glass and
stone. The surface would be chameleon-like in its ability to absorb the same
amount of heat as the surrounding rock and soils. Or it could reflect more
heat. Foundations and floors would be made of concrete. The arcology
is designed for 250,000 human residents. Perhaps as many as 40 percent
would be transient during the summer. The arcology would have a relatively
small footprint at 25 million square feet (575 acres), with 40 stories, 3 of
which would be underground in the gravel and bedrock. The arcology
would share design motifs with other pyramids. The base square is 5000 by
5000 feet.

The maximum height of the building would be in the range of 1000
feet, divided into 100 stories. The average floor area is 8,000,000 square
feet, with a bottom floor area of 25,000,000 sq. ft. and a total floor area of
832,500,000 sq. ft. (and volume of 8.3 trillion ft^3). Density would be 435–
521 people per acre (the denser parts of Hong Kong are 1000 and Mumbai
are 791; efficient urban density is estimated at 100-500 households per acre
or up to 1200 people per acre). The feel of density is reduced with generous
wild and agricultural spaces.

Over 30 percent of interior space is designed for housing, forming neighborhoods within floors. An additional 18% is commercial space

including retail stores, offices, restaurants, grocery stores, and production facilities. Additional office space may be developed in the center as part of the housing area. About 5% is reserved for administrative or civic use, distribution centers and convention facilities. And, another 5% houses utilities and construction yards at eight levels. Green space for food growing and parks covers 36% of the total area, divided equally, mostly on the lower levels. These areas include various environmental uses and provide open, public streets and parks. Another 16% serves as public circulation services and cultural zones comprised by schools, medical facilities, day care centers, theaters and art facilities. And, 10% of the area allocated for cultural zones is underground and specifically designated as a Virtual Reality recreational park (similar to what Soleri proposed for his Hyper Building).

The main terminal on the southwest side of the structure would connect the arcology to the local port. MagLev lines would take travelers from the port and from the main station to local stations within the building. Inside the building, travelers, residents, employees, and visitors would shift from mass transit to local elevators. These would primarily operate within public spaces throughout the interior. Within vertical neighborhoods, vertical circulation would be enhanced by extensive use of escalators or elevators. Pedestrian movement is aided by escalators, diagonal or spiral escalators, moving sidewalks, elevators, electric conveyances, and bicycles. Because moving sidewalks expend relatively much energy, they would be limited to short stretches.

The pyramid has a relatively large surface area, compared to a sphere, for its enclosed volume. In this location, the larger surface area is necessary to capture solar energy. The shape may also be kinder for presenting openings to light and air. Although there are seasonal variations, especially regarding light, the overall climate is polar.

Residences would be modular polyhedra with curved ceilings inside; curved spaces may have psychosomatic effects on the residents. Pedestrian distances are measured by walks and in minutes. Other wheeled vehicles, such as bicycles or battery-powered carts, could be available for delivery or recreation. The arcology could have a proper mix of handicraft labor and intermediate technology, but the main work would be service and support.

The arcology would require more participation. Furthermore, the design of public areas should contribute to a more democratic style. It would be a governable size, with clearly defined boundaries. The arcology would specifically work to maintain a framework to represent the people, as well as the other beings in the ecological system, by reemphasizing goals. Efficiency would be less important than participation in good judgments.

The principles of an arcology, from using marginal land and imploding spread to habitat density to internalizing and miniaturizing the urban effect and linking the habitat, energy and environment, serve as a model for a stable, exciting construct. The creation of a world arcology as the living headquarters of the United Nations on a separate continent is an important step for representing humanity in one intense, iconic building.

Figure 144. Antarctica World Pyramid Arcology (Home of the UN).

Experiment: *Is the City a Trap We Cannot Escape?*

Is the city a human ideal? An environmental ideal? The city can offer ideal environments, as well as different kinds of physical environments, from streets, and squares, to religious monuments and parks. What is it about a city that commands awe? First cities were of bricks and concrete. John Thackera thinks that cities are held together now by human attention spans, which may be more gaslike than solid. The technology that dominates attention, such as the wireless infrastructure, just adds a new layer, as people still live in brickworks. Certainly, the city has always been an incubator of new forms and ideas. And, the medium for cultural transmission has been ideas, more than genes or bodies—and, there may even be standard units of cultural transmission by imitation, such as memes, Richard Dawkin's phrase for the unit, of which cities, agriculture, and fashions are examples.

Human groups have adapted to their environments, have changed them and have been changed by them. Some human groups were trapped by their environments and could not change. Sometimes the adaptations themselves can act as traps that limit human behavior.

Are we urban by nature? Have we changed from hunting to urbaning? Of course, we may be preadapted for cities. We are clever social animals. We prefer edge habitat so as to move between other habitats. Urbanization is a characteristic of an edge species. Civilization produces edges that people like. Wilderness is fragmented into islands and patches. So, there is no more deepness, no more interior to wilderness. And, there are more edge species: Raccoons, coyotes, crows, and rats. Are humans an edge species?

Is the city a trap then? Is the city a sink? Of course, at some point, cities are no longer adaptive to the environment. What is the trigger, if it is a trap? Size? The degree of aggregation, as well as overall density, which results in population growth and survival, varies with species and conditions; undercrowding, as well as overcrowding may be limiting. Aggregation can enhance group survival. Fish in a school may tolerate higher doses of poison than individuals. Bees in hives can create more heat than individuals. Applied to humans in cities, aggregation is beneficial, up to a point (bee or termite colonies that get too big divide or die).

For a long time cities required larger numbers of people to be intense enough for stimulation and creativity. A growing population creates environmental resistance to itself, in the form of reduction in the reproduction rate as the population approaches carrying capacity. Due to various time lags, e.g., to increase when conditions favor or to react to unfavorable crowding, the density can overshoot the capacity. Human population is controlled to some extent by self-crowding. The overshoot has been on a local scale, never on a global scale. The Russian geochemist S. Vernadsky concluded that the property of maximum expansion is inherent in living matter as it is for gas expansion or heat distribution (and ideas in cities?). The pressure of life can be measured in terms of velocity. For cities, immigration can increase fast and cause overshoot and subsequent oscillations. But, what if all cities were to overshoot simultaneously?

Perhaps the trigger is the number and pattern of connections in the city. Increasing populations can lead to intensification of production, through labor or mechanization. Interdependence becomes overconnected and then a trap. How important is connectivity? Cities are unavoidably entangled in global nets, now. Cities used to be limited by the local carrying capacity. With global nets, self-correcting feedback can take too long. Localization makes feedback visible and more immediate. It could make more self-reliant cities.

Perhaps the trigger is the speed of change in the city. We are changing conditions so fast that new adaptation is necessary to fit into conditions we cause through growth and adaptability. What is the solution? Plan or adapt? Violence might work. War has been a way to get out of traps. War can also destroy filters. But, there are other ways to change filters—by perception or thought. War can destroy cultures and ecosystems, as well as cities, so it may be too destructive to work as a solution.

Perhaps the trigger is alienation or abstraction. Humans have always worked to abstract their specialness. The city seems to be another myth of separateness and independence from nature. The city is seen as a laboratory of human creativity, kept apart from the mother of nature. Lewis Mumford saw city life as a compromise between the hunting stage and the farming stage, where the female principle of home is wed to the male principle of predation. Perhaps the city is a throwback to hunting, when people were more likely to wander territories, and were less attached to a place.

Assuming the city is a trap, how to we get out? Growth is a way to get out of some traps, but sometimes it leads to a larger trap. It can be broken, in species as well as corporations, by diversifying—and how does one diversify? Find the differences in the environment through education, specifically through ecolacy. Perhaps education can be designed to afford understanding of adaptations and traps. Another way is to legally level the playing field with antitrust laws. One solution to growth is mature development within limits of age and size.

Traps can be escaped by reformulating goals, or by weakening or straightening feedback loops, by adding new feedback loops, or by recognizing them or altering the structures. The way out of the trap of escalation is to avoid getting in it, refuse to compete, negotiate, or interrupt. This is where education can offer alternatives, through the filters of numeracy and literacy. The solution to cheating is to redesign rules in a direction to achieve the original purpose of the rules. In fact, design and redesign are ways to remove traps.

Or is escape impossible? The city is a different kind of trap, that offers intensity and opportunity, but requires massive imports of supplies to survive. The size and scale of cities create the dual centers of attraction and despair. Maybe we are addicted to the trap.

Are cities the ultimate creation of civilization where people can enjoy culture free from want or physical extremes? Or, are cities a gross alteration of nature that destroys human life and dignity—a terrible ecological error. As long as cities grow without negative feedback, the second will occur. When ecology helps the city ecosystem fit in the surrounding systems, then the achievement is worthwhile. The city depends on its environment.

Humans developed as generalists, using creativity and flexibility to cope with unstable conditions following the glacial climate of the last ice age. If the future returns to the climatic variability of earlier interglacial times, this variability would impose difficult conditions on agriculture and cities, which developed in the relatively mild 5,000 years past. The most recent trends to interdependence, just-in-time supplies, and globalization are not adaptive to any instability in the system. Arcologies might be viable alternatives then.

Experiment: *Can We Escape Other Traps?*

Cause and effect are hard to separate. The same environment that challenges a culture with some kind of change, also offers opportunities. New resources can stimulate economic activity and increase the level of living in an expanding cycle of production and waste. But, cycles that do not operate with the right kind of feedback function as traps. Thus, on an elemental level, phosphorus becomes trapped in an ocean sink, and can only be recycled by long geological processes or by specific harvests.

A trap is a metaphor, since we depend on nature and society as a foundation for life. Traps function in different ways. The use of resources by a people, where the replenishment rate is constant and the rate of use exceeds it, is a serial trap. This trap results in ecosystem degradation that is less reversible. The industrial form mistakes the rate of discovery for the rate of recovery.

Agriculture is an energy trap, because it allows a higher concentration of energy in higher yields, but then it requires more energy be put into the system to maintain it. The system has to produce more energy than it uses to be sustainable, with a surplus for trade. As the population increases it requires more food and energy, and more land to provide it. Eventually, people can not go back to simpler forms of life, because there is no territory left. People are tied to a particular place and have to communicate to adjust to sharing places.

Many other situations can act as traps. Other traps include: Addiction, policy resistance, arms races, and the tragedy of commons. People become addicted to new commodities that cities offer. Addiction is a trap, since it requires higher rates to appease the need. Addiction can appear in human social systems as well, such as dependence on government subsidies or reliance on chemical fertilizers to improve crop yields. Addictions, such as those to foods or oil or money, make it difficult to escape from a trap, a trap being a kind of energy well or gravity well. Addictions can amplify some emotions, such as fear or hate, especially as they relate to the threat of an end to addiction. Addictions can justify illegal behavior, especially those that seem necessary to continue the addiction. Of course, many cultures are addicted to the illusion of control and power. The US is trapped in the belief that it is the most powerful nation and a necessary peace-maker. Eventually the trap is escaped, or more likely destroyed, or collapses with its victims.

Global capitalism can lead to a consumption trap. Capitalism claims to serve the wants of the people, but it spends half its income creating more wants in people. Not many of those wants are real, or as real as cereal and roofs. Few of the soft services satisfy real psychological needs. Markets advance individual desires and not social goals, by offering running shoes, not inner city restoration. Instead of being free from economic want to develop their potential as creative human beings, people are trapped in a consumer cycle. Self-actualization is postponed for self-gratification. Furthermore, capitalism can undermine traditional cultures by offering consumerism in the place of guides for behavior. Social roles seem irrelevant by comparison, if the good life can be bought without effort.

Being in a trap means much-reduced flexibility and fewer choices. Being in a trap makes one vulnerable to many other changes. When the weather got colder, then hunters and gatherers could move south. Cities could not. Climate change can drown whoever is in the trap. Civilizations are more fragile and more vulnerable to smaller climactic changes. Paul Shepard suggests that the entire Neolithic revolution has trapped us in behaviors that only end in madness. The feedback seems inevitable. Can we change and get better? Can we take the trap apart? Violence is one way to break out of a trap. Madness another. Ecological design is a better alternative.

Experiment: *Could Arcologies Replace Cities*

At present over three billion people live in cities, about half of the total world population of 6.3014 billion (1 September 2003), most of them within 100 miles of the ocean. At the same time, cities close to the ocean are being threatened with storm surges and tidal inundation; there is talk of protecting them and moving them at great expense. What if, as a thought experiment, we built new arcologies on the least productive lands at higher elevations, and moved 3 billion people into them? Essentially all urban people would live in arcologies, except for some small traditional communities living in wild ecosystems or agroecosystems that had to be managed. Currently, the expansion of urban areas with roads, power grids, and other infrastructure, is interfering with the basic functioning of many ecosystems. Modification, conversion and destruction of ecosystems disrupts the complex interactions within and between ecosystems, the hydrology, soil structure, topography, and vegetation; it changes the complement of species, and it causes a loss of diversity. The new replacement 'domesticated' systems are simpler, less mature, and less diverse.

We have achieved great horizontal growth, much like a fungus. However, if we want to be like a smarter fungus, slime molds for instance, we need to learn to cooperate to grow up and become more dense. The larger metropolitan regions are covering wild ecosystems and agricultural fields with single-family houses, malls, building, recreational areas, and roads—all of which are car-centric or auto-morphic. This means that energy and goods are also spread thin. Such systems of things are hard to control, hard to keep safe, and hard to remain interesting. It might be worthwhile to compare human systems to mature ecosystems; we are creating pioneer individuals that do not live well in concentrations. We are creating edge individuals and not those who can live in interiors and share resources, or can develop new resources with cleverness and intelligence. City designs do exist, however, which incorporate the properties of mature systems, as well as the characteristics of ecological thinking.

An arcology, as defined by Paolo Soleri, is a city that embodies the fusion of architecture with ecology. The arcology concept proposes a highly integrated and compact three-dimensional urban form that enables radical conservation of land, energy and resources. Arcology eliminates the automobile from within the city, and with it, the fifty percent of land devoted to automotive needs. The multi-use nature of arcology design would put living, working and public spaces within easy reach of each other and walking, supplemented by elevators and airport things, would become the main form of transportation within the city. An arcology would use passive solar architectural techniques such as the apse effect, greenhouse architecture and garment architecture to reduce the energy usage of the city, especially in terms of heating, lighting and cooling.

The small footprint of an arcology, combined with many built-in gardens, would allow rural space and agricultural fields to be closer to the city, and a part of the immediate urban environment. Wilderness, also, would be much closer to population centers in arcologies. Psychologically, the intelligent design would be more conducive to inspired living, the kind found in traditional culturally-significant cities at certain times. The proximity of agriculture and wilderness would allow people to participate more, with the full range of benefits that comes from growing and cultivating plants, as well as being able to immerse in the otherness of the wild.

The sizes of arcologies range from 250,000 to almost a million people, although smaller or larger ones are possible. For the sake of argument, assume that the average arcology is the size of Soleri's proposed Novanoah, at 400,000 people. At that size, it would take 15,734 arcologies to house the planetary population; this number is less than the number of cities in the United States in 2004, at 19,354. Assuming that the area under the arcology is about 5 square kilometers (almost two square miles), the surface area taken up by arcologies would only be 78,768 square kilometers, which is only 0.00054 percent of the land area of the planet—that is half of one thousandth of one percent—149.45 billion square kilometers, or roughly 0.0167 percent of the land area currently under concrete and asphalt, which is 4.71 million square kilometers or 3.15 percent of the land area of the planet.

The shapes of arcologies could be as diverse as any. They could be large pyramids, filled with living and working spaces, connected by transits and illuminated by light wells. The traditional ziggurat modernized would offer a good ratio of sunlight and truck gardens to size. Arcologies could be empty tubular pyramids with modular dymaxion attachments, that could be moved between arcologies or new sites. They could fit the shape of the land-scape, as does the Palouse Arcology, mostly underground. Arcologies could be built around small mountains, in bridges crossing canyons, or threading through coastal seas.

Since Soleri's heroic designs, arcologies have been confined to com-puter games and have become elements in science fiction and cyberpunk films. With more prototypes being designed, one structure may be built in the next twenty to thirty years. A proposed project for Tokyo Bay, the Shimizu TRY 2004 Mega-City Pyramid, would be the largest artificial structure on the planet, at 2004 meters tall housing 750,000 people. The external structure would be an open network of megatrusses, supporting struts made from carbon nanotubes to allow it to stand against high winds, earthquakes, and tsunamis. Separate buildings for housing and offices would be suspended from the supporting structure with nanotube cables.

It seems that arcologies should be part of a whole package of chang-es, brought about by ecological planning on a global scale. The experiment would not require that arcologies replace archaic populations living in hu-man-modified ecosystems, or even all low-density habitations or traditional cities. But, they could be new cities situated in infertile areas.

Many cultures could live in optimum configurations in their territo-ries, as part of wild and domestic landscapes. But, we also need heroic ar-chitecture. Heroic design and extravagance in life is needed in general. It is not contradictory or antithetical to frugal lifestyles or to restoring a healthy environment. Life is exuberant; energy is used, lives are lived and used, not wasted or saved. Life is the accumulation of individual experiences that are remembered by subsequent generations.

Thought experiments about arcologies appear throughout this work. The best response to a question about what would happen as a result of some actions under some circumstances may be a thought experiment. Through that, we can create designs for arcologies and discover answers in a dialogue with others.

Figure 150. The Lean Linear Arcology (Credit: Paolo Soleri).

Experiment: *Rethinking Cities*

Cities developed as adaptations to changing climatic conditions, a place where food could be stored against droughts and hard times. Then cities started to change mobility and habits, as well as work styles and possessions. But, what is a city? A city is a complex structure characterized by buildings, roads, residents, density, division of labor, air-conditioning, domestication, partnerships, patterns of movement, concentrations, intensity, and miniaturization. Termite mounds have all those characteristics, also, by the way. We need to think about cities as a form of ecosystem that can be made more self-sustaining and complex.

The city ecosystem, that replaced a native ecosystem, has been greatly simplified; there are fewer plant and animal species, and lower diversity, leading to homogenization, especially when cosmopolitan or favorite species are present. Plant productivity declines, due to fewer plants and lower rates of photosynthesis. Connectivity to the surrounding matrix is lowered—the matrix itself is divided by increased patches and corridors, such as roads, which perforate the landscape and also act as barriers to plant and animal movement. The number of patches increases, which decreases interior species and increases edge species, such as squirrels, raccoons and coyotes. Native patches and corridors, such as woodlots or streams, are reduced or eliminated. The agriculture around cities is more homogeneous, with decreased fallow areas. Stream corridors are degraded or destroyed, making a city more vulnerable to floods. Sewer systems route more water faster to an ultimate sink, lakes or the ocean. Household wastes affect the landscape directly, if they are buried or burned. Nutrient and mineral cycles can be disrupted. Local weather can be disturbed.

Many new cities are based on the prevailing assumptions of old cities, such as having as much energy as possible and using as much land as is wanted, for buildings, roads and parks. Many of the new towns, particularly in England and the United States, were created with automobile-oriented layouts. Roads, for instance, were designed and built with no knowledge of road ecology or the native ecology. Many of these cities were planned for an optimum population within their area, but today are proudly said to be 'still growing!' Growth is a problem for most cities, especially those that seem idyllic with many amenities and opportunities for work or pleasure. Unfortunately, by not limiting growth, those cities attract so many new residents the infrastructure starts to fail.

Cities use too much energy, which changes their own microclimate, with heat islands and dust domes. The masses of materials used to build are used inefficiently. The energy to run the city is greatly concentrated—thousands of times more than in a native ecosystem—and the energy is used inefficiently. Unless the infrastructure, from buildings and power lines to sewers and roads, is kept up, it can decay and cause greater inefficiencies and problems. A budget crisis, from loss of tax revenue or improper use of funds, can affect the infrastructure and most amenities of a city. The disruption of supplies, due to social and political actions, such strikes or attacks, can lead to other problems, from displacement to disease. Even in small, well-managed cities, there has been an increase in violence, sickness, drug use, crime, and impoverishment, all of which require some response from the political structure of a city. Ian McHarg noted that with crowding, there was an increase in mental and physical illness from stress.

Because cities are so concentrated, and occupy a relatively circumscribed area, they require distant support systems for food, materials, and energy, as well as a smaller system for pets (monkeys, cats, canaries, or fish) and support machines, such as automobiles. This requires constantly increasing inputs and outputs.

No one seems to have determined how large should a city be or if a certain density is needed to inspire creativity and invention. For instance, in neighborhoods, the goal of separation resulted from older traditional ethnic neighborhoods with some crowding. Later neighborhoods lost the feel of community, as people had their own quarter-acres and personal cars to reach the local malls and supermarket complexes. They felt more isolated and more vulnerable to fear and violence, especially the middle classes. Many of these people decided to intentionally bring back the sense of community as well as reduce waste and vulnerability to violence by creating intentional communities, such as ecovillages. But, even these goals did not always have a density that inspired creativity and invention.

Goals are obviously imperfect. There is always some element missing that results in a suboptimal condition. And, the resulting dissonance prompts a response, which may result in an improvement or a less satisfactory element. Perhaps with continued experiments on ways of living, combined with an ecological perspective and forms of ecological design, communities can optimize human happiness or at least satisfy it minimally. Much design has to do with solving one problem or a set.

Very rarely is a city redesigned to solve problems. And, very few cities are designed to avoid recognized problems. We are not sure whether to design for maximum numbers, maximum luxuries, or maximum happiness. Based on our knowledge of systems and maxima, it is dangerous to design for maxima of any kind, or even for optima—perhaps it is best to aim for a satisfactory level.

The properties of good places could be used to help guide designs for cities. Action, for instance has to be appropriate for the scale of the city as well as the environment. Because of the complexity of interrelationships, the program of actions has to be coordinated. Ecological designs become actions to create good places, such as urban places, within a good environment modified by good cultural practices. Design needs to reflect that uniqueness in all its productions, to insure diversity at the urban level. Diversity can lead to richness, which is also an important quality of good places. Good urban places offer the circumstances necessary for conviviality, that is, for all residents and beings to live together. The consistency of a place permits residents and others to anticipate changes and disturbances. When this happens, health becomes more of an attribute of living in that place. Design needs to address health above all else at the level of good places and good cities, although we can also design for safety and peace.

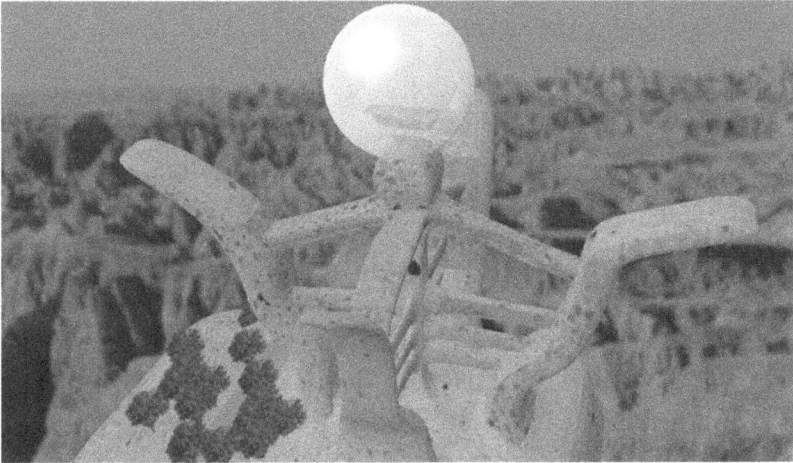

Figure 153. Rethinking Cities: Sunflower Arcology (SD, US).

Experiment: *Is the Future a City Planet?*

Why do not we see the global scale of operation? Why does it seem that our human imagination about space is kept down to the size of a city? Is the space of the world the same as the scope of globalization?

The image of the city world predates the advent of globalization but it is not yet visible. The classic architectural styles aspire to a certain sameness or universality, regardless of national boundaries. It is a city world of course, but also is forced and assumed. H.G. Wells thought that civilization would be led by technocrats and operated as one entity, the city, as a city world. Science fiction continues to reimagine the world as a single structure, a single city.

Megalopolis is a constant image in literature and popular magazines, where cities grow and connect to create reasonable bands of urbanization, with expanding communication and transportation. This was magnified to the scale of the world by Constantin Doxiadis. His approach to urban planning was anticipatory and remedial. His contemporaries included Yelena Friedman and Buckminster Fuller. Friedman scaffolded a parallel floating world on top of the ground-bound one. Fuller expected technology to continue building the earth and into interplanetary space.

Should we want there to be one city on the planet, or at the extreme one planet city? By matching the world as one entity we overlook the fund of previous imaginations. Recently the qualities of connectedness, continuity and sameness have changed from becoming wish images to being projected outcomes of development.

One city, one basket, one pool, one ecosystem, one planet. One global scale that we have yet to grasp and understand. Driven by constant growth. I think what the architects forget is that sameness and repetitiveness can be avoided by addressing the locale, rather than relying on constant growth.

The idea of a planet city is not compatible with the ideas of arcology, which stress ecological limitation of the city to a unique landscape, as well as the reduction of the human population and impacts to fit those limits. Soleri noted once that arcology is anti-colonial and anti-Empire.

With our panoply of advanced communications and virtual worlds, there is no reason to cover every continent with concrete. We now have the capacity to map the world beyond a simple global city model. We can map the world's resources as well as our greatest desires and dreams, and then situate them in arcologies, cities, towns, and villages that encourage dwelling that fits in local places woven with civility, wildness and otherness.

Figure 154. Abandoned slum in Kowloon.

Part 11: Sharing Things—Economics

Experiment: Why Tolerate The Rich? Can We Eat Them?

In hunting and gathering societies, in which humanity lived for 50,000 years or more, most everyone was equal, although a few displayed more talent for hunting, healing or peace-making (so we forced them to be leaders). With the advent of agriculture, there were large surpluses of food, and some more aggressive people claimed much of that for themselves. This accidental happenstance has been perpetuated for the past 9000 years without much questioning or protest. The suddenly rich in barley were able to set up a tradition of a few rich overlords.

But, why should we respect this? Why do we allow the rich? What is the role of the rich? Is this just an accepted part of a curve of differences? Sometimes the curve seems very steep, but we still allow it. Now, the situation is one of extreme disparity. Maybe it continues, because everyone dreams of becoming rich themselves. Is it because richness is too much a goal? Financial richness, I mean.

In 1996 the UNDP estimated that the wealth of the world's 358 billionaires exceeded the combined income of countries holding 45 percent of the world's people. And the gap grows. There were 793 billionaires in 2006, worth $2.6 trillion USD. If we added millionaires, what would that be? In 2004, there were 7.7 million millionaires, with $28.8 trillion. That $31.4 trillion was more than the entire world economy, at only $25 trillion. Possibly billionaires and millionaires hold over 90% of all human material wealth. These want us to maintain the current kakistocracy.

What is the value of having rich people? As models of happiness or moral behavior? Not likely. Even science confirms that money does not guarantee happiness. And, the real level of moral behavior fills many news tabloids. As role models or leaders? Some wealthy people have made good leaders, but many more poor people have been that. The rich seem most comfortable in their roles as rich people. As beneficial contributors? Some rich people have been very generous, but not most. As economic engines? Studies show that there is almost no trickle-down; few rich people start companies and employ people, and most are content to compound their richness through savings (preferably off-shore and untaxable).

So why do we have the rich? Habit? Fear? Envy? Maybe we should become mature now and do without them, at least without the extreme disparity that divides humanity now. This does not mean that some people cannot have 4 suits of clothing or 5 kinds of tools, just that having 6 houses or 10 cars or 3 islands is simply not acceptable, especially while millions of people get sick, starve or die of preventable ills or diseases. Historically, after some cities collapsed, some of the fallible rich were killed and eaten.

Experiment: *Understanding & Solving Inequity Now!*

The inequities between human beings are mathematically large and seemingly impossible to achieve without luck and long intergenerational control. Inequities tie up money, the symbol of reward for 'spending' time working. That is bad economically for any system, because then the symbols do not flow or work. Inequity limits many people to basic housing and often inadequate food. That is bad because then people are tired and unhappy and have less to look forward to. But not having enough also deprives those people of 10–30 percent of the years of their lives. People without a safety net or adequate money die younger, after having more trouble with sickness.

Yet, those with billions hidden away or invested in business still are capable of making themselves unhappy, and although they do live slightly longer, it is not appreciably more than the average of all people. So, taking money out of circulation does not give them any significant advantages. Why wouldn't they reduce the stress of having by sharing?

We could blame inequity on early historical patterns that became habits, then traditions, and finally rights to keep more. Hunters had experience with extra food or diamonds. They did not turn that extra 'bling' into kingdoms. What if first civilizations had shared better, like hunter tribes? Why didn't they? Do people always divide things unequally if there is a surplus. Did the Mesopotamians, or Chinese? Yes.

Did the Kwakwaka'wakw? Yes, apparently. With a uniform distribution of fish, through fluctuations one region may suddenly have more fish, an instability permitting a sudden symmetry-breaking amplification to occur. As difference increases, positive feedback becomes stronger until one region has all the fish or gold. Although it may shift to another group again.

Is there a way to reduce the differences in wealth? Or even guarantee a near complete equity. Would taxation work? Probably not as long as the oligarchy could control tax laws. Increase the minimum wage to $100 an hour? Probably not. Increase human consciousness of the costs of disparities? How would that be done? Perhaps by law, require people with home larger than 1000 square feet to accept from 1 to 70 homeless people (depending on each increase in size of 500 square feet).

Maybe we should just try to pool all global money in one pot and divide it equally between everybody. Why not? The logic is irrefutable. If everyone were rich, minimally, then there would be less unhappiness and crime. The economy would be smoother, eventually, and there would be less cheating. Of course, people would make mistakes and lose much of their money and it would start to concentrate in the hands of others.

Maybe the governments should keep it and just give out allowances every month.

Experiment: *Restore Safety Nets for the Disadvantaged!*
In a typical city, some people will be sleeping over subway gratings, in church doorways, and in parks and yards. Others will be warehoused in large rooms filled with small cots. Still others will be sharing apartments or houses with relatives. Some people will not be able to afford enough good food, or any food. Although some sick people will be treated in hospitals and charity wards, others will only be cared for by relatives. And, others will be getting no care, and a significant number of them will get worse and die, often alone. Some people will not be able to move to better jobs or opportunities. With the economy being 'disturbed' by hurricanes of greed and crime, many people will be robbed of their savings and investments.

At one time in the US, the government established a pattern of safety nets, so that people who lost jobs would have support for a period of time to get new ones. People who were poor or generous throughout their lives would have some income when they were old. The economy was managed so that rent and food prices did not fly out of reach.

What happened? Apparently some rich people revised the laws to reduce or remove any government help for its people. This is reminiscent of the old scuffle in England in the dark 1800s to remove help for the poor because they were simply losers in the competition of evolution. Herbert Spencer thought the weak should be eliminated, so he opposed poor laws, charity, sanitation, education, clean water, and pure food, among other things. Is the US so much less enlightened, now?

Please, restore our safety nets—all of them. What should these safety nets be? To start, (1) Social Security and Medicare for everyone.

Then (2) a federal infrastructure work program, like Civilian Conservation Corps in the 1930s; perhaps it could be called the Infrastructure Restoration Corps or something. The nation is falling apart because of the false economy of cutting most of the budget except for the rich and military. The possibility of jobs exceeds the labor force.

New laws (3) to control corporations and banks, not just their profits and bases but their charters; banks would guarantee all deposits (if not by holdings then with the salaries of the managers). Corporations, have been taking over investments and government services for a profit (the rights to do so are given away by lawmakers for consideration).

New buildings (4) need to be constructed, with all levels of rent opportunities. Homes need to be taxed by size, to encourage smaller, more efficient ones. Food banks (5). Social health services (6) for women and families. Educational help (7) from kindergarten through college (either subsidies or free) for anyone who wants it.

These things could be a start. The government, which people support with taxes, has to provide these nets. And, it has to change the taxes to ensure that it can provide them for generations.

Experiment: *Are There Enough Jobs?*

There is no shortage of work. There is no shortage of money. So, why
are so many people around the planet jobless? Of course, some people
make their own work, from an idea or from understanding how to produce
something with more quality and efficiency. But there are still hundreds
of millions jobless in Asia, Africa, Europe, and South America (and 25
million in the United States; that number includes those who have given up
searching or are not counted properly as unemployed).

There are much needed repairs on the infrastructure of our
civilization: Buildings, houses, streets, bridges, and utilities, the pipes,
conduits, canals, wires, motors, generators, and all the other equipment
for supplying people with clean air and water, electricity, and devices
for communicating. Most of these things need repairing, upgrading or
installing. But, there are not enough salaried workers and professionals to
perform the needed work.

New skills and workers are needed to address developing problems.
In the transition to alternative energy, more jobs will be developed.
Another necessary kind of work is cleanup, from regular garbage pickup
and disposal to the collection of vast plastic islands in the ocean. With
continued climate change, entire cities will need to be protected from rising
water levels or even dismantled and moved inland. An explosive population
requires more and more services. Education needs to be increased in every
country. Many people, perhaps 40% of the planet's population, have
discovered this; unable to get salaried jobs, they rebuild computers or collect
and resell fabrics. In doing so, they are able to survive at a minimal level.

There is no shortage of money. The US, for instance, has set aside
vast sums for that country's protection from real but minor threats, with
ultra-sophisticated weapons. If the US were to reduce its military to a
reasonable defense, then the remaining 80% of that could support at least
5 million jobs, which would lead to a better security than weapons of mass
destruction. Other Nations, including North Korea, Russia, Israel, and Iran,
could in a similar way direct military spending to jobs needed for real food
and energy security, as well as education and an adequate infrastructure.
Tax reform in many countries would add uncollected billions with taxes on
bads and on heroic inequity.

In one sense, we need to structure these jobs in ways that people can
have living wages working only 20 hours for 3–4 day weeks. Many cities and
countries have already pioneered the ideas of flexible work: Half-time jobs,
because half-time workers are more productive and efficient in their jobs
(according to US government surveys); shared jobs where two employees
share one full-time job (40 hrs/wk); working groups, where a number of
workers are paid to produce a large item, such as an automobile, rocket,
or airplane. Two things especially improve moral and productivity—

management attention to worker needs and worker empowerment in design and decision-making. By recognizing that there is plenty of work and money and ideas, everyone can support themselves.

Experiment: *We Could Limit Work & Leisure*

With a global unemployment rate of 40%, it is obvious that capitalism has not worked. Of those who do work, perhaps 50% or more have their own business or service. It is obvious that corporations have failed to hire people, because that might interfere with the economic bottom line: Always make a profit, exactly or higher than predicted. Fewer than 20% of communities offer special breaks to small businesses, the same breaks that they woo corporations with. It is obvious that most communities are deficient in options. Nations have failed to offer meaningful employment. Corporations have failed to offer living wages. Communities have failed to support workers.

Leisure should mean freedom from work, but not all work. The garbage needs to be collected and recycled. Otherwise, without work as a focus, most people are useless. Because their dreams, their lives, are small, unformed? Not so much that, as they were never taught to find meaning in some activity.

At the same time, we need to avoid the world of leisure. Except for the gifted, always a small percentage, the history of those who do not have to work is sad. Especially the landed gentry or the self-important class. Only vice seems to relieve their boredom.

Work has to be tied to meaningful production at least, and at best to the health of people and their environments. The ecological bottom line has been ignored; the environment was labeled as a free good, and now it has been degraded. Eventually, production will not be possible at all, although in economic terms, the final write-off will still benefit the corporation for one last gasp. The first work has to be restoration of the environment.

Adjusting to the difficult changes that are necessary to survive the looming emergencies, seems to be a good time to start offering work to anyone who wants it. And what is more meaningful than saving lives or cities or ecosystems? The transition to a balanced planetary system is going to take the efforts of billions of people over a generation or two.

Studies support the cases where people are given 4-8 weeks per year vacation, as well as 3-6 months maternity and paternity leaves; it recognizes their needs and contributes to their happiness. People with part-time jobs test happier than those who work full-time. People who share jobs are more enthusiastic than those with separate, individual jobs. We should promote work as to what is necessary to keep civilization going. The same with vacations.

Experiment: *What Would Happen If We Ignored Efficiency?*

Efficiency has been enshrined on a million factory walls. It is now so critical and important that it has replaced production as the goal of those factories. Factories, which now layoff most of the workers in the name of efficiency, and pay somewhat less millions for sophisticated machines, and become less efficient in the large sense, which is employing people in a community to provide necessities and luxuries of that community.

The name used to be jobs, for healthy communities but those sloppy social glues are just left to the families, now, so that politicians and community leaders can better plan a better, more efficient community, perhaps inhabited by the efficient machines that got the jobs (and are often located overseas). Sure works out that way in cartoon movies, like *Robots*.

But is efficiency that important? Nature does not seem efficient, yet it is very productive. Recently biologists started rating the efficiency of ecosystems and asking if more efficient systems could compete better. Other biologists noticed that efficiency alone cut back on diversity and flexibility, which are necessary to long-term survival, which meant that the efficient systems failed first, in the event of a novel disturbance. Alas, nature is full of novel disturbances, which may be why efficiency has such limited value in evolution.

What if we ignored efficiency in production, not just in industrial processes, but in our lives? Efficiency is not always a good. Efficiency has unified the world for travel, but it is monotonous and boring, to go from the same room to the same room in the same seat. There seems to be more uniformity, as production becomes more efficient, with fewer workers, and styles are shaped by the lust for efficiency. Maybe more differences are needed. We should want more diversity in everything, not just for aesthetic appeal, but for redundancy and balance.

Under some circumstances, efficiency is a good thing, but too much of it could lead to ill health, or a flood of identical objects. Very few people value or collect sheets of toilet paper or small plastic GI Joes. If everyone were a street sweeper, happiness might be rare.

People want individual things, rare things, or beautiful things. They want things that are precious, or different or owned by other famous people. But, these things are not produced efficiently. Value and rarity in durable goods seem to be more desirable. Efficiency is most valuable itself as a way to produce the basics of food, shelter, and personal goods for everyone.

De-enshrining efficiency, and paying more attention to the labor force would increase productivity. A number of studies in the US in the 1930s and 1940s found that the best way to increase efficiency was to keep asking the workers what to do to improve their conditions—that improved productivity more than better lighting or new machines. Factories in Scandinavia, mid-Europe and parts of Asia have improved worker conditions, rather than laying them off or shipping the production overseas, and found that they are just as competitive in the world market (for example Volkswagen and Volvo). Things that are well-made may be more expensive at first, but demand for them would bring the prices down a little, as other companies try to compete to make them. And, all this could be done without sacrificing jobs and communities, in the name of an ideal that we pretend is as concrete as food.

Experiment: *Hollow Myths & Permanent Bubbles*

The symbols and myths, like freedom and equality, are a shell for America the nation. The nation has been high-jacked by the oligarchy—the economic elite, the religious elite, the corporate elite, and the military elite. The small, ridiculously wealthy and overrewarded, overprivileged group that steals in the name of patriotism, that cheats the poor in the name of democracy, that plunders the holdings of banks in the name of trickle-down sharing. That recklessly borrows hundreds of billions of dollars in the name of fiscal responsibility for stocks and jobs. That destroys the manufacturing base in the name of profit, that—oh, hell, they just betray, lie, raise fears, and pretend it's for the common good.

The shell is a façade of ethics and power that no longer exists. The bubbles are reflated to allow quick profits for some, but the values have been lost. We revel in the illusion of wealth, not realizing the stark reality that we are really poor and bankrupt. We are comforted by another illusion, that we can become wealthy, all of us. Sudden consciousness of the disconnection is going to cause massive suffering. The absurd fantasies have been a trap.

The dominant global economic paradigm has been a mega-Ponzi scheme. It is predicated on nonstop, infinite growth. It will collapse, but by then there will be no graceful way to collapse. And, we cannot predict when it will collapse. Probably after some investors panic at the realization of what it really is. Who knows when that will happen? Tomorrow? Next year?

Moral decay is reflected in the physical decay of the environment. Cities are plagued by breakdowns, most often of water systems. Thousands of jobs are moved to cheaper labor nations in the name of 'Free-market' and 'Fair trade.' Corporate totalitarianism will enslave many in dead-end jobs. Soon, we may be rioting in the streets, without enough money or food.

The global superpower that we are touted to be will decline. The glorious tomorrow will be just another dingy day looking for work, maybe for one of the few, unskilled service jobs left. Some of the elite will promise a miraculous resurgence. Many of us will never reach even close to the standard of living promised by the elite demagogues to secure our vote. Let's vote them out and start over with a leadership draft or something better.

Experiment: *Why is Industrial Civilization a Problem?*

Our modern problems reflect an unbalanced and immature image of the earth, the earth as a machine, for instance. People sometimes constructed their worlds from preconceived notions, and many of these worlds did not survive, because they could not adapt to the environment. Our modern cultures are defective for this reason. The modern attitude toward nature as a resource has resulted in pollution and depletion of resources. It has allowed humans to overpopulate their habitats. Recent productivity studies indicate that the optimum sustainable human population is far below the current world population.

Even worse, decisions regarding resources are still made exclusively on short-term economic rationalizations and lead to material shortages and environmental degradation. The crises of environmental degradations are crises of cultures. Monocultures of the industrial kind lead to 'dedifferentiation,' that is, the decomposition and destabilization of complex structures. A species or culture that destabilizes its ecosystem through misbehavior risks its own extinction. Human beings make changes to ecosystems that endanger themselves and risk extinction.

Industrial society is constantly mobilized for emergencies, in battles against ignorance, poverty, diseases, and terrorism. Industrial development has never been nonviolent or respectful to people. Industrial production has its own unique style, shape, and scale.

Is the industrial city a jungle? Herbert Spencer saw the world as a jungle, where life was nasty, brutish and short. When Spencer coined the phrase 'survival of the fitter,' he based it on the competition in industry, where entrepreneurs fought for money and power. Karl Marx considered the industrial system a cage. Perhaps rioting is people lashing out at their cage, when people cannot leave cities that do not meet their needs.

One problem in industrial culture is the production of flatscapes. Our attempts at improvements have proceeded without order, without sufficient insight and perspective, without sufficient confidence, without a comprehensive plan, and without a great dream. Our politics has been corrupted by special interests. The structure of our civilization comes from anonymous builders, mediocre designers, minimal engineers, and rapacious financiers.

We work within the rules as they have been for decades, rejecting any alternatives as too utopian. The rules themselves have been shaped by centuries of social metaphors and utopian ideals. They do not exist in place, either a human place or an ultrahuman place. They are designed to be no-place, without problems, poverty, weeds, storms, or hard ground. Because they are nowhere, flatscapes, like demented unplanned utopias, lack any reference point.

Speed, so beloved by industrial cultures, can be a problem in archaic cultures. The Kelantese people in the Malay consider haste a breach of etiquette and ethics; slowness is important. Mental illness can be caused by the increased tempo of life, probably in any culture.

Information pervades society, but it is for an information market. Between the bulk of raw data, uncoordinated information and the acceleration of activity, the social fabric gets torn and fragmented. Projects are fragments of work. Industrial information work is bad for physical and psychological health, from bad physical conditions to high pressure and low control. The information economy can lead to disconnection, loss of identity, and loneliness.

Physical laws create patterns in space, as well as in human history. There are simple kinds of patterns. A linear pattern tends to be interpreted as progress or regress. This is the dominant concept of modern history: Unending progress. Despite the chaos of individual events, there does seem to be a direction of gradual improvement. Marquis de Condorcet suggested that civilization will always move in a desirable direction. People become frustrated when it does not, and there are emigrations, destruction, collapses, and revolutions. A second Jeffersonian revolution is overdue.

Experiment: *Stop Stealing! Theft Is A Problem*

The utilitarian aim of greatest good for greatest number has been vulgarized to mean the greatest number of goods for those who can afford them. In our attempts to manufacture the good life for those who can afford to buy it, we have stolen clean air and water, quiet nights, darkness, open spaces, and other indefinable qualities. Soils are destroyed, wildlife is killed. We devour nature to assuage our disease; we try to fill our emptiness with goods. We can only gain past a certain point before our gain causes the loss of something else that we need to be healthy. Everything we do is a kind of theft, although we deny it.

Modern technological society ravishes nature and mutilates humanity with the products of its materialism. Industrialization has distorted people's lives and cheated them of bread and justice. In a mass consumption society, people impoverish themselves spiritually while impoverishing others materially. This is theft. As with the Christian Ten Commandments, most loss can be reduced to theft, whether of a life, mate or name. Most of our modern problems can be considered the consequences of theft, such as of life, common sense, and finally choice.

Over the past two centuries, industrialized countries thieved great quantities of raw materials to create luxuries. Then they disseminated the ideas of wealth, equality, opportunity, and indulgence to many countries without industrial opportunities. There can be no peaceful future for civilization when such disparities, and popular knowledge of them, exist. The cult of competitive consumption seems to be the universal solvent of the modern world. Everyone wants what some have. The industrialization

of Asian nations is seen as a solution to shortages of manufactured consumer goods, although fiscal imbalance and pollution are down-played.
Even worse is the unavoidable waste. Probably over fifty percent of the
productive effort in United States goes into making things which contribute
nothing to the material standard of living.

Consumer desires must be satisfied promptly or despondence
results. Believing the line of succession ends with claiming his inheritance,
the industrial consumer feels no obligation to provide for the next generation. The consumer justifies narcissism as a preliminary condition in the
search for consciousness, truth and morality. Consumers become embroiled
in their own causes, locked in the idealism of adolescence, and repudiating
the lessons of history. Hegel noted that the greatest lesson of history is that
nobody ever learns the lessons of history. And, Cicero wrote that those
who do not know the past are like children. Americans and people of other
nations seem intent on validating these insights. Like children.

This repudiation of history creates a prevailing historical amnesia.
Consumers retain little more than a dim notion of the past. Universities
report a lack of interest in events that occurred before the current year's
athletic season. The sense of time falls in upon itself, collapsing like an
accordion into the present. Knowing nothing of history and expecting
nothing of the future, people cannot escape the fearful isolation of the
present. They join together in a melancholy herd, clutching at everything,
stealing things, but holding nothing fast.

Without the depth of history, experience is shallow and short, and
intelligence is thin. Technology has reduced the globe to a single, closed
system, which humans can share according to their financial powers.
Our direct experience of the world has become shallow, in spite of faster
travel. Travel used to broaden the mind, but now it narrows it. We travel
in sealed corridors like boxed goods, comforted by homogenized foods
and the English language. Our cultural adaptations to the pressure of
homogenization throttles individuals and groups.

Our 'overweening bumptiousness,' which the Greeks called
hubris, lets us behave as though we were too privileged to be members
of the earth's ecological community. We name things and dismiss them.
We draw lines around them to separate ourselves. We build our own
environment, mathematical and sanitary.

John Fowles observes that most of us remain firmly medieval and
distanced from what we cannot own or fully control. We assess most of
nature as what is not clearly for us must be against us. We cannot accept
indifference, the nonhumanity of nature. We seem incapable of realizing
that the destruction of the Amazon, much as we deplore it in its remoteness,
is our responsibility; we consume the materials from tropical forests. Our
growing emotional and intellectual detachment is the greatest threat to

nature. Heroic narcissism has replaced nature with humanity; nature no longer provides the mirror to reflect human aspirations—a television screen does. Narcissism is a threat to nature and humanity.

We have allowed the thefts of life, intelligence, identity, and choice to continue. We kill millions of domestic species every year and many millions of wild species by accident or by design. We steal from animals and plants, from the earth, and from our own descendants. With the theft of common sense, civilization makes decisions that are destructive to most individuals, as well as maladaptive for societies under ecological restraints. The theft of choice occurs when we are dominated by the ideas of industry, which form the outlines of a tragic world view: The primacy of humanity, the supercesion of the individual, the achievement of happiness through the accumulation of things, and the perception of the incompetence of nature. These ideas are false, and programs that assume them for conditions are doomed to fail eventually, perhaps destroying the very things we need.

Experiment: *Reestablishing Economics—Koinomics*

Economics is the way of managing the place of a human culture. Traditional styles of economics historically began with (1) Subsistence household reciprocity, within bands of people. As groups got larger, the (2) Distribution of resources and goods was extended within tribes. Later, and with a change in scale, the way became (3) Redistribution through a more formal process; virtually all economies from chiefdoms to tributary, socialist, and market economies are redistribution types. Then, with a nascent globalization, the modern economy developed a way of (4) 'Pluridistribution,' based on more abstract accumulation; this severely linear economy is dominated by different ideas of ownership, capital, corporate independence, and capital conversion and erasure (Wittbecker 1976, *Eutopias*). However, the myths, problems, and massive negative effects, such as the destruction of local and global commons, of this abstract system are being addressed and fixed by (5) 'Koinomics' (as an extension of ecological economics). Theodore Roszak called for a nobler economics, one not afraid to discuss spirit, conscience, moral purpose, and the meaning of life. He, with Schumacher, Boulding, Daly, Cobb, and others, grasped that economics has to be treated as a subdiscipline of ecology.

Koinomics requires the equal apportionment of 'resources' to all living, interacting participants in the global or local commons. This involves recognizing the entire legacy of the planet as it is created, developed and maintained by its tenants, and allowing reasonable access by nonhuman beings for their needs, since we ultimately depends on their 'services.' For human beings, it means limiting our interference with living webs and biogeochemical cycles. This way is supported by the 'rule' of all beings (see

'Panocracy'), although in the human legal system, humans represent the interests of all other beings, much as they are starting to do now. This way is enhanced by the wisdom of harmony (see 'Harmosophy') and the drawing of and by the making of ecological zones, which emphasizes the relative separation of wild and artificial areas.

The reapportionment of 'resources,' claimed and unclaimed, would be a radical and large-scale undertaking under the auspices of ecological design. We can make ecological designs, mostly to restrain humanity and restore balances of systems, but also to coevolve with the development of natural systems. We can create a framework, with clearly delineated functions at each level from personal to the planet, with the integrated approach of koinomics. And, we can advertise this to everyone so they can participate. Then we, they, all of us can say what else should be done.

Koinomics requires a change in the operating rules—the principles, standards and practices—to allow profit only after natural systems and human needs have been restored to health and protected. Resources rights are distributed to all beings equally, before human beings can exploit theirs for their own needs. Profit is a recent idea that arose from a system concerned with raising capital for exploration and trade. As a critical factor of modern economics, it has become a trap, from riding a sudden exponential expansion based on fossil fuel use, to the complete indifference to environmental costs and human suffering.

A koinomic model shows how human welfare can be increased without uncontrolled growth or excessive profit, using rules that limit drawdown, overshoot, or any of the other catastrophic trends that are deepening. It would slow resource depletion while increasing productivity and efficiency. It would reconnect processes into cycles of reuse, and lower waste streams causing pollution and dead-end sinks.

More than just the making of basic goods, koinomics can provide a spectrum of services, while protecting the sources of energy, materials, and labor. It can make the strong connections to ecosystem and community health, to the availability of good jobs and the security of homes and work places. With strong connections to education and healthcare, with limits to salary differences, the model would provide an ethical responsible approach that could lower poverty rates, unemployment rates, and all forms of sickness. By understanding the limits of ecological and political systems, by respecting the properties of healthy ecosystems, place and cultures, it could still provide for the needs and luxuries of most all of humanity through mature development and reasonable takings.

Koinomics could be effective by being responsible for educating all people to feel their connections to their place, because, until they feel them, they will not act ethically or ecologically. For educating people to realize that long-term health and sustainability requires healthy places.

Part 12: Sharing Power—Politics

Experiment: *Ecologists Want to Rule the Earth?*

I had trouble recognizing Richard Watson's characterization of Deep Ecology in the winter 1985 issue of *Whole Earth Review*. He seems to have quoted the program correctly, but has drawn the wrong conclusions from it.

For example, he seems to find somewhere in the literature of deep ecology that the ideal human impact would be on the level of hunters and gatherers (which is Paul Shepard's ideal anyway), with a global population of five million (after quoting Naess's ideal of 200 million). However, limiting human impact is not the same as returning to hunting or gathering or to subsistence agriculture, which no deep ecologist has ever recommended. Watson presents his own ideal at 500 million, based on a concept of cultural flowering. That figure is the same as Daniel Kozlovsky arrived at intuitively and close to the one I calculated for an optimum population based on global net ecosystem productivity (455 million, 1983)—an odd surprise.

Watson's criticism of ecological ethics is based on a confusion of value. A nonanthropocentric ethics argues that beings have self-value, although not necessarily human value. Saying that the generic term value is independent from 'man' as Watson does, is meaningless. An ecological ethics can address the limited relationships of all beings without becoming entangled in the fuddle of reciprocity or sentience. By basing an ethics on 'what is,' deep ecology avoids the 'absolute' that bothers Watson.

Deep ecology is not anti-anthropocentric. In fact, it accepts the necessity of an anthropomorphic, anthropocentric, and anthropometric logic. It does not accept the extremes of such a logic, however, which assigns to humanity all value and creativity. Furthermore, deep ecologists do not claim, as Watson accuses, to know what is right or good for humanity, let alone the ultrahumanity on which we all depend. Indeed, most deep ecologists urge caution and noninterference with primary cultures or wilderness areas.

I do not know what the ecosystem wants Watson to do, as he begs it to tell him. Does anyone else know? But I doubt if it wants him to die for the human species, as he complains. Deep ecology is not concerned with the good of the human species apart from the diversity of species and the health of the ecosystems in which we live.

When apocalyptic rhetoric is not heeded, human societies vanish, leaving behind their monuments and deserts; hundreds of human cultures have 'bit the dust' in the last 3000 years alone—keeping archaeologists in business. Contrary to what Watson would have us believe, deep ecology is not concerned with telling people what to do or with saying what is right. Moreover, it stresses that we should not always attempt to say what is right for all.

Watson's fear for the ambition of deep ecologists is unfounded. It is unlikely that Naess, Hardin, or Skolimowski want to 'rule humanity' (as Watson warns us), although it would not hurt to keep an eye on them.

Experiment: *Shredding Corporations*

The growth of corporations caused a growth in government in response,
mostly to help those hurt by large corporations: Small farmers, small
businesses, and individuals. According to Robert Reich, the government
expanded its constitutional powers, not by amending the constitution, which
would be the appropriate way, but by allowing the Supreme Court to give
broader interpretation to the Constitution. The government ended up trying
to regulate parts of the economy or those parts that corporations were
unwilling to perform. At least until government simply became a willing
partner in the changes. Economic government now dominates public
government, which helps it.

Individuals cannot, and corporations will not, take responsibility
for public services such as highways, airports, national defense, national
parks, schools. So, government takes sole responsibility (as it should be
responsible for national things). National defense for instance, has enriched
many corporations, but the corporations do not even provide efficient or
competent work, in many cases. Public resources such as oil and forests flow
to corporations as subsidies. Radio and television licenses were distributed
to corporations. Now, corporations dominate public airways. Public lands
used to be the support for individuals. Now, they support corporations.

As private corporations became larger and international in scope,
local and state governments have been unable to regulate their activities.
That would take a strong local or national government or international
body. In their own interest corporations work to reverse the intent of the
people. For instance, although U.S. President William Clinton in 1992
promised a pro-employment policy, the minimum wage was not enacted,
and interest rates were raised. What the people voted for was reversed
by the invisible management of corporate government. The legal system
becomes another tool for power, to justify the distribution of wealth, as a
result of favors. This destroys the neutrality, fairness and dignity of the law.

Under corporate tyranny, urban parks and wilderness suffer neglect
or destruction; parks and forests used to be the symbols of freedom. The
free market did not decide that; corporate managers did. The current
economic style is too great, fast and reckless for ecological systems to absorb
its impacts. The scale of things is an independent problem that can ruin the
best intentions of policy.

The broad and deep middle class of the U.S. in the 1950s acted to
stabilize society. It also pulled up many from the lower class. Now, the new
structure of the work place, with its steep hierarchy in megaglobal corporations, is destroying the middle class, with help from banks and corporate
relocations. Security is lost; community is destroyed. Corporations control
the workplace without interference by government, unions or competition.
The lower class deepens with little outward flow. The bottom is a trap now.

Economists try to bronze the economy in its current structure; but it is a changing system. Since it is changing, strategies that are appropriate at one stage are totally inappropriate at another—this is the remorseless working of tragedy, where successful strategies are applied in inappropriate circumstances.

Unfortunately, the modern system privatizes the gain and externalizes the loss to the commons, considered as a pool of "unowned resources," where in traditional societies, it was surrounded by rules for use. As long as this is possible, it is profitable to charge the cost to the environment. Externalizing costs works fine in an uncrowded world, where the costs are negligible and can be absorbed by natural processes. Resources were traditionally seen as free for the getting; air, water and land were seen as free sinks.

Labor and polarization have forced corporations to liquidate capital assets, such as old- growth forests, rather than keep them as long-term capital. The answer is provided by Marvin Harris, in his review of recent economic history: After the second great depression starting in 1932–3, the U.S. government tried to flatten out the boom and bust cycle of our capitalist economy by manipulating taxes, interest rates, and money supply, by subsidizing jobs, and through corporate welfare. It worked for a while.

As the automated production of goods became more efficient, goods-producing jobs declined. On the other hand a large, well-educated labor force, mostly women, was available at relatively cheap salaries. Information processing and human services proved to be far less efficient than automation. Harris suggests that corporate bureaucracies wasted labor and lowered productivity faster than automation could save or raise it. Corporations were no longer efficient enough to expand production out of sale-generated income and took on additional debt.

Corporations have gradually increased their dependence on deficit financing to supply capital. Harris notes that corporate debt has gone up fourteen times while federal government debt has gone up only three times, as a percentage of gross national product. As more money was owed, more was paid to service the debt and less was available for cash flow. More short-term debts were taken on to keep up cash flow. Corporations can borrow faster and pay more to borrow, but can keep raising prices to consumers, so they can keep borrowing—or liquidate some of their assets, which is probably why Burlington Northern and others allow massive clearcuts: Cash flow needs due to long-term inefficiencies and massive short-term debts. The impulse to save a corporation may override the need to save the natural heritages, such as old growth forests. Unless, we can remake corporations into service entities, again. We have to.

Experiment: *Incorporating Regions: A Palouse Example*

In early civilizations, the advancement of the state was expected to contribute to the welfare of its people. Corporations are recent devices created by states for public purposes. Most early American corporations, for example, were concerned with travel (turnpikes and inland waterways) or safety (fire insurance)—they resembled public agencies more than profit-seeking associations. The exclusive privileges and political power granted to corporations were based on the implicit promise of social services.

The association of economic development with national wealth allowed incorporation laws to be broadened. The corporation was given the constitutional rights of an individual. A corporation is a legal entity, independent from its founders, with its own rights, privileges, and liabilities. It is, however, required to obey laws and pay taxes; and it is accountable for its deeds in courts of law.

Unfortunately, as private good became identified with public good, corporations became larger, more acquisitive, and less concerned with social services. The quest for profit now has the effect of violating social amenities, such as clean air and clean water, instead of ensuring them. No responsibility is taken for environmental degradation since no right of contract or fair use of property has been breached. However, there is no reason that entire ecological regions could not be incorporated.

One way to represent all voices and interests in a region would be to incorporate it. The region incorporated would focus on a core business: To ensure the integrity and continuity of life and all its connections and to secure the opportunity for development free from undue interference. It would operate to optimize values, like any good corporation, but the values would be ecosystem values (fungus values and earthworm values, as well as human values).

The important advantages to incorporating a region are the same as for incorporating a business. (1) Managerial flexibility: The stockholders are separate from managers; responsibilities are assigned by needs of the corporation. (2) Limited liability: The corporation borrows and repays. It shields its members from hazards to which they would otherwise be exposed. (3) Financial advantage: The ownership of assets can benefit stockholders and the corporation. (4) Tax advantage: Investments in the good of the corporation may not be taxed by any nonlocal government. (5) Estate planning and longevity: The corporation exists indefinitely beyond the lives of its participants. (6) Central management and representation: A large and complex business needs operational and managerial efficiency. Many of the participants have no direct voice in the operation—they must be represented.

A temporary Board of Directors would adopt bylaws, elect working officers, approve stock certificates, open accounts, and arrange a stockholders meeting. The stockholders would elect new directors, possibly from local representatives or directly from elections, and decide on dividend declarations.

Stockholders, as citizens of the area, would turn over common and state (or regional) property to the regional corporation, for example, the Palouse Corporation, which would issue stock certificates to the stockholders. The corporation would allocate the purchase price of stock to capital at par value. Most of the shares—the percentage to be determined by the board as necessary to the operation of ecosystems—would be treasury shares. Anything more than par value would go to capital surplus, and only capital surplus could be distributed as dividends. Stockholders have the right to receive these dividends equitably, without resort to traditional distributions of wealth.

Stock certificates denote ownership of the corporation. Although the stockholders own the corporation, they do not own the property of the corporation, the entire region, which is owned by the corporation itself. Stockholders, as individuals, communities, or cities, could make agreements about how business would be conducted, about what resources would be used or traded.

The elected board of directors would make decisions of distribution and limitation. Percentages would be deducted from the interest for the operation of the corporation and for equitable distribution to areas less favored by chance with biological or geological wealth. Furthermore, since the dividends would be distributed among people according to net ecosystem productivity and resource availability, no advantage would be gained by areas having large populations.

The basic functioning system would be considered capital, thus limiting the amount of human use of resources and probably the size of human populations. Interest would accrue in the form of net ecosystem productivity and diverted percentages of materials, such as gold or water.

The Palouse Incorporated would solve the problem of having to value ecosystems in monetary or quantifiable terms; its systems would be untouchable capital. The human value of resources like copper or water would be equated to the technological cost of recycling or producing them.

Raw material and energy are only two facets of the capital of a corporation—another is human ingenuity and the production of cultural capital. Thus, human wealth would not be limited by restrictions on the availability of resources, but rather by a shortage of ingenuity.

An incorporated region would be instrumental in conditioning international corporations to their social responsibility and in internalizing all costs. This corporation and governments could use traditional means, such as credit access, low interest rates, and setting priorities on equity issues, to evoke public interest in smaller and healthier human endeavors.

Experiment: *Is the Cost of Civilization Too High?*

What is the relationship between levels of complexity and energy and information? In an ecosystem, the energy required to maintain the system is inversely related to its complexity; succession decreases the flow of energy per unit biomass until the system reaches maturity (R. Margalef 's concept of maturity). In a mature forest, for example, almost 100% of the energy is required to maintain the state of the forest. Any system formed by reproducing and interacting organisms develops an assemblage in which the production of entropy per unit of information is minimized, that is, waste is minimized in cycles.

The same relation seems to apply to cities or civilizations. The simplest economic transactions were between individuals who gathered food or made tools and then traded. The number of artifacts seems to increase as the populations increase. The number of artifacts seems to increase as the complexity of a culture increases. Materials can be used to express power. With the increase in specialization and complexity, came individual traders, then guilds, and finally corporations. A complex (or mature) civilization tends to use 100% of its energy to maintain itself.

In every case where civilization has become more complex, according to Joseph Tainter, the cost of maintaining that structure has required an ever-higher percentage of income be set aside for maintenance; that income is no longer available to increase the standards of living. Modern civilizations have avoided collapse only by overusing fossil fuels to hide or override their true costs.

The ecological design of civilization has to address the true costs of maintaining levels of control and luxury (and some level of luxury is a necessity). Ecological design has to be the creative modification of ecosystems to repair or enhance their ability at self-organization and the maintenance of their complexity and diversity. But, most of all, it has to reduce some forms of complexity and improper uses of energy. Design has to fit human exploitation within the limits of a changing planet. The human population with its expenditures and levels of luxury, has to fit in the sum of ecosystems providing limited 'services.'

Figure 172.
Ruin of Ur.

Experiment: *How to Elect Leaders & Representatives*

Leadership is needed to avoid collapse. Most people seeking office seem to be seeking fame or a cash flow. The interests of the people, and the nation and its living territory, are being neglected. This is bad for many nations. A few government systems have selected good leaders, but the success rate is low. Some candidates buy their way into power with combinations of money, fame, advertising and lies. A fair number of leaders currently are being indicted for illegal activities in office. There has to be a better way. Perhaps suggested by past experience or a rational approach.

Way Zero: Things stay the same—But the process would be limited. The nature of the political activity would change. For some offices, such as ombudsman, people might have to be drafted. Campaign spending would be limited to an arbitrary amount, such as $50,000, all of which would be provided by the government from voluntary gifts or by small taxes. Public services would be required to give each candidate 100 hours of television ads or appeals (as part of their bandwidth licensing).

The offices themselves would be staggered and different lengths of service. For instance, presidents would get one term of five years, senators 4 years, House of Representatives, 2 years, Other elected offices, such as ministers or members, would follow this pattern. No one could serve more than one term for each office. Although voters could inform politicians, lobbyists would be forbidden. This would require radical transparency, in everything, not just the true costs and sources of cars or hamburgers, but of all public proceedings.

Way One: The Chinese Method—Take an Open Exam. First, a candidate would take a written or aural test on national history and politics with some cultural and ecological content. Second, she would have to pass a debate on 3 popular issues, such as employment, immigration or legal punishment. Third, he would have to create a sample plan for dealing with a hostile nation or creating an educational program on languages. Then, these three steps would be judged and graded by a panel of impartial experts. Only by passing, could a candidate run for office.

Way Two: We Got Politicians—TV Show with Judges & Audience Voting. Candidates would first offer volunteer Videotape applications, to a council of judges, possibly Nobel award winners or race car drivers, who would select them for the show. During the show, they would engage in a free-form conversation. People would add input in a tree-like network of posts; each would have a single issue or argument. People would cast votes on-line. The winner would begin serving in office.

Way Three: People would nominate others to serve for a particular office. Anyone could be nominated, from car salesmen to gurus. The candidates would be screened a government agency, and three would be chosen to compete. Another committee would select the winner.

Given the fact that the characteristics that make people lust for leadership often detracts from their abilities to lead, perhaps we need a form of draft to select leaders, or perhaps we need to consider persons who have been groomed for decades or from birth, like the Dalai Lama.

We should begin by examining different models of power and authority: Autocrat, guru, revolutionary, or transformer, as W.I. Thompson suggests. The transformer has authority but no power; the autocrat has both; the revolutionary has power but no authority; and the guru has neither. Perhaps we should consider the guru, since wisdom is more important than power or authority. Leadership requires trust and informed judgment. These are the qualities we need, not unfounded confidence, brashness or simply being rich.

Making these changes occur all at once would seem unrealistically disruptive, but would it be? Would it be more disruptive than a Senator's impeachment or war, than runaway inflation or economic collapse? Would people's lives change that much? Would it be worse than losing a job or having a business fail? Speaking from experience, I say, "No."

The UN might be required to help autocratic states. There will be problems with any changes in political processes, but as long as people have an interest and can participate in the process, things should improve. Eventually, or rapidly.

Experiment: *Supplementing Democracy*
With the collapse of Communism as a major form of government, everyone has been jumping on the bandwagon of democracy, without critically appraising the suitability of democracy for many other cultures in the world. It is possible, maybe likely, that democracy is not the best form of government for many cultures. Democracy has obvious weaknesses that make it an unsuitable political system for many cultures. James Madison argued that a democracy got weaker as it became larger and suffered from the violence of its factions; he suggested that a republic, by contrast, was more suitable to govern large populations, as its structure let it become stronger as it became larger. And, so the US has a representational democracy in a republic. But, it seems as the republic has gotten larger, it has become weaker and more violent.

Quite possibly democracy is too weak in an ecological age that requires a system sensitive to complex connections and finite limits. We may need a form of government that uses an ecological perspective, regarding the role of humanity in trying to manage the entire planet. But, what would it look like?

Most books present a limited number of systems of government. For example: Representative, participatory democracy; Direct, representative democracy; Totalitarianism; Communism; Monarchy; Dictatorship; Republic; and, Anarchy, where any minority is not bound by majority. Similar to an anarchy, it is possible to have rule by consensus instead of a majority. Of course it is possible to have a democracy without elections—a demarchy. And, in a republic, it is possible to have a supranational rule, where states vote by population.

In a eutopian framework of independent cultures, coordinated by the UN, it would be possible for any culture to have rule by traditional means. For example, the Tukano may wish to be ruled by an Elder. Other groups, like the Iroquois, may choose to be ruled by a Council of Elders elected by female participants. Many cultures have been ruled by elders; these have minor differences in selection, since the ruler may simply be the eldest person or may have the deepest ancestry, as in the Kwakwaka'wakw. An intentional culture may choose to be ruled by a council of ecologists or artists.

Any kind of rule has to address the same problems: How to provide equality, peace, and security for the members of the culture. That means being able to disarm militancy or level wealth in cases of extreme luck or access to resources. Rule also has to allow for a smooth transition between generations. How this is done is more important than cleaving to a broken, romantic system, whether communism or democracy. This kind of experimentation may be best done in relatively small sizes where failure can be tolerated and change can be made quickly.

We might be too self-centered, and we might be thinking too small. The planet is occupied by thousands of different systems and billions of participants—that all should be represented in our system. To recognize the rights of plants and animals, and their habitats, we might need to create true panocracies, that is, a rule of all, in the interests of all.

Figure 175. Democracy in Action.

Experiment: *Redesigning Politics as a Panocracy*

For Aristotle, politics was the science of the possible. The city (or *polis*)
was a human artifact whose structure could be modified by reason; it was
potentially a work of art in which only the capability of the artists limited
the expression. The city was made for the amateur; and it produced
more complete men (women were erroneously considered lesser beings
at the time). The city was necessary for politics then, as face-to-face
communication. Now, communities are different, larger and dispersed, so
politics has to change. As a science, politics (or ecocybernics, a neologism
meaning governing the house) has been concerned with two things basically,
a way of distributing power and luxuries internally within society, and a
way of surviving contact with other societies.

The function of politics in general is to ensure that decisions are
taken at the right level. A state protects individual freedoms, guards national culture (values and identity), and holds internal groups accountable for
use of power. Communities and individuals also have to make decisions. In
addition to having rights, citizens also have responsibilities to participate in
government and to live as wisely as possible, to make good places. Often,
duties have to be made known through education and communication from
the government.

Ultimately, politics is about the definition of reality itself. Social reality is created. Different societies have had different realities. Politics is also
the art of creating new possibilities for human progress.

The current system is defective, however. Although it is admirable
to work within the system to prevent further environmental degradation, it
might be necessary to produce a change in consciousness that would lead
to a new political paradigm. That paradigm has to be the rule of all, for all.
That means every species and every place—a *panocracy*.

In a panocracy humans must represent other species, who have a
stake in the rule; humans already successfully represent the young, very
old, mentally handicapped, and legal constructs such as corporations,
who cannot represent themselves. So, it is no leap for some humans to
represent other species or places. It would be like a modern totemism, for
representing a silent majority.

A panocracy would recognize that ecological, economic, social, and
religious phenomena are part of the broad definition of politics. The basic
goal of politics is the survival of the community. As survival is survival
within nature, politics rests on an ecological foundation. The organization
of a community must be in accord with natural laws. Political participation
depends on accurate information, much of which is provided by ecologists
as well as engineers, sociologists, and participants.

Politics occurs in an ecological context. There can be no separation
of politics and ecology. Every political act has ecological consequences,

and every ecological decision is a political demand for control over use of the environment. Ecological consciousness must be identified with political consciousness. Politicians need to think about the hunger and squalor of billions of human beings and the destruction of habitats with billions of ambihuman lives, before concentrating on missiles and private fortunes.

A developing politics could be derived from our ecological identity, but that identity may not be comprehensive enough to be effective in politics. Maybe our perceived dominance over nature will be a problem that politics cannot solve. Certainly ecological politics is going to be more complex than traditional politics, which was after all designed for the human city. Perhaps we could create an open, self-conscious ecological politics.

For this politics to be global, it would have to address all cultures and all interests, human and nonhuman. A panocracy, a 'rule' of all beings, like a democracy, could result from a formal legal system, where humans represent all interests. A global ecological politics would have far more restraints on it, and limits to it, from the complexity of the emergent global connections and structures. But, in being comprehensive, it would leave nothing out of consideration.

Figure 177. A Rule of All Beings would include this vole & coyote (Credit: Defenders of Wildlife).

Experiment: *The Next American Revolution*

Why has someone not formed the Next Jeffersonian Revolution Party? That should have happened by the mid–1800s. Obviously our institutions and governments are senile, and in need of some good changes.

What can we do? Form the party! That should be a better name than my own self-started party, the Practical Radical Ecology Action Party. And, we need to do it. Our own nation, and others, and the planet, are being beset by disasters and catastrophes. This is an Emergency, or a perfect storm of emergencies converging from the fog of long-term, invisible, slow-acting, large-scale, fuzzy trends—gigatrends— that we seem unable to anticipate until they start to bite us.

Why can't we see it? Because it is slow (remember the high-school experiment, kettle of water, with frog, slowly boiling), because it is long, longer than our limited lifetimes of 70^+ years, and much longer than our normal horizons of 2 to 20 years. Because it is invisible, we cannot see how the form is developing or how it could represent a danger to business as usual. Because it is accelerating; we tend to underestimate or overestimate the speed of reactions in natural or global systems. And, possibly because it has not been intended or designed by us.

And, it threatens our current serious way of life, the small victories we have made for ourselves in an industrial trap. The current political parties are narrowly focused on power and votes (and rich benefits for the elected). Even the 30 or so new parties like the Green Party seem immune to the building dangers. None of them, *none* of them in the US seem to be aware that we will soon be facing emergencies as they hit us, one after another, or all at once. Please help form a viable party to address the most important issue since independence.

Figure 178. We need a revolution to escape our traps (Credit: Google).

Part 13: Computers & Communications

Experiment: *Why I Avoid Fast Things*

Civilization, with its growth and speed of exchange, has been accelerating for many generations. Many cultures in the past operated at an optimum speed for their circumstances. But, now a large industrial culture wants to operate at a maximum speed, for competitive advantage they say. It should be a wild ride. No brakes, no steering, no windshield apparently, but a big rearview mirror, to see those 'losers' in the dust.

And, speed affects everyone. Alvin Toffler concludes that a great technological and cultural wall will separate the slow from the fast, making problems for joint ventures. The slow will be left behind as the fast produce their heads off. But, what are the products of these ventures? The debris of advertising fads, such as mink toilet seats, or the tools of real needs, such as evaporative water purifiers? Or real durable goods like wooden furniture or houses? Economics will decide, then people can choose from the limited array of junk.

By submitting ourselves to positive accelerating feedback loops in economics, we distance ourselves from such primary meanings. Nature possesses power that is not speed dependent. That power has history and momentum. Human consciousness has already had a revolution—from the wild to the tame—and we regret it. Animals used to be directly experienced; now, they are humanized and domesticated. Any revolution to the "fast" is bound to have negative consequences. The acceleration of culture makes accumulation of tradition more difficult. We tend to think the present is over and has no hold on us. We become obsessed with newness and youth because they hold our attention briefly. Consumption contributes to the growth that speed requires for its launches, but it leads to self-destruction.

Some people have begun to respond with a desire for slowness. There are movements for slow food to counter fast food, for slow transportation in travel, to reap benefits unknown by fast travelers. There are slow media and slow celebrations. These are important trends (not gigatrends, but mega-trends). Milan Kundera notes that remembering calls for slowness.

I remember living in a culture that operated on solar time in a predominately natural context. I think that is an appropriate speed for most human things. In art, for instance, or clothing repair or logic, one should take the time to pull the threads into place. I also remember sky-diving, speeding in a sports car and sending book files to China by computer, but these are exceptions to the more comfortable and productive pace of life.

Experiment: *Can We be Wise in the Age of Computing?*

To drive civilization and manage the planet, we have assumed that we
have enough, data, information and facts. All we have to do is apply them,
although that may take wisdom. Is wisdom something we can buy or get? Is
it innate or do we have to learn more?

It starts with knowledge. Gregory Bateson defines wisdom as
knowledge of the larger interactive system. Wisdom is a perception of
relationships and relativity, an awareness of the wholeness of things without
losing sight of the unique particularities. It joins the left and right brains in
a union of logic, poetry and feeling. It reintegrates knowledge with values.
It implies making judgments in advance, infusing elements of older wisdom
into a new expression. The wisdom of cultures forms a perennial philosophy
of the human race.

Jonas Salk defined wisdom as the art of disciplined use of
imagination in respect to alternatives, exercised at the right time and in
the right measure. Judgment is required as to what is right, and judgment
may be an innate art. Wisdom consists in the exercise of judgment, based
on qualitative criteria, in conflicting situations. Facts are judgmental. As
Garrett Hardin notes, without wisdom, compassion and empathy will
destroy the object of attention, not improve its circumstance. Humanity can
affect the elements of individual, society, spirit, ecology, and political life, by
affecting them all together.

Wisdom can be thought of as a new kind of fitness, supplanting
the biological kind of evolution. It is a cultural thing. Humans have made
radically different conditions that they must now accommodate. If the
mind is exposed to economy of nature, as revealed through living systems,
humans will recognize the necessity of balancing values. Total win-lose
conflicts are unwise. Value systems concerned with dynamic equilibrium,
aesthetics, complementarity, reciprocity, justice, interdependence,
adaptation, reconciliation, and intuition, are the language that biology
speaks.

Myths (with transformations) and metaphors (with structure of
integrated differences) are modes for conveying ecological wisdom; they are
less concerned with survival than the survival value of a good fit between
the dualisms of life. Equilibrium is needed between self-restraint and self-
expression, between self-protection and self-restriction. Not self-expression
or self-restraint, or exponential growth or some plateau, but all in the finest
fit. Fitness attunes us to limits. Wisdom cannot be dependent on perfect
knowledge; it does not exist. Even in ignorance, we must act "as if" we were
wise, circumspectly, with caution and respect.

Experiment: *Is the Planet a Hard Drive ?*

There is an article in the August 2015 *Scientific American* by C. A. Hidalgo, "Thought Experiment:. Planet Hard Drive." It is an interesting article, in spite of its misdirection. However, the planet is not a hard drive. Likewise life, also attached to that metaphor, is not a hard drive. Life needs space to expand, much as atoms incorporate space around their structure, to fill out its form. In-*form*-ation is order. Humans increase information as order in landscapes and buildings, and in the planet as natural capital.

What does it mean to calculate the potential information storage capacity of the earth? Since entropy grows with order, eventually there will be no more motion and change, only forms that cannot even degrade. Change now is mostly neutral and uncoordinated, as it is in human actions and productions.

With every atom in the universe as a computer (not external surely), we could store 10^{90} bits. M. Hilbert and P. Lopez calculated that humanity stored by 2007 about 2 times 10^{21} bits. The earth could store 10^{56} bits. Most information on earth, however, is in biomass, about 10^{44} bits.

Hidalgo states that earth's hard drive is fuller than ever. But, this is wrong; it was fuller 70 million years ago and at the first radiation of body forms. He further says order on earth is rare. This is not so; the order may be chaotic, but it is easily visible in the patterns of connections. Does it have machine-like order? No, but so what?

Hidalgo states that order emerges and persists in the universe due to 3 tricks. The first is the *flow* of energy downhill. A draining bath tub lets order materialize in a whirlpool. For enduring order, we need the second trick of *solids*. DNA preserves order for a long time, as does a stone pyramid. Without solids, order passes too quickly. To explain the emergence of complex forms like a forest or a city, we need a third trick, the capacity of matter to *compute*. This last would be true if to compute meant to reproduce to fit with an environment, instead of calculate and process instructions. Of course calculating and processing instructions are two things that living organisms do to survive in an environment.

Hidalgo uses the example of a tree computing, but the tree is not a machine. It decides when to withdraw sap from insect pests or to share sap with injured trees nearby. The tree does create order, but the purpose is to survive and do well.

The author claims that imaginary orders give us access to other knowledge. True, and it also lets us generate knowledge. But, then he claims that the economic market makes us richer and wiser. Richer than what? Wiser than who? Wouldn't experience growing corn also do that?

We agree that the ability of human networks to create order or information is constrained by historical distribution, language barriers and isolation, but we overcome that with trade and communication in larger networks, although often we neglect making goals or act directionless.

Hidalgo predicts technology will be a globalizing force, pushing a merger between bodies and information machines—that might make things worse, socially. Then, he concludes that parochial institutions such as religion or patriotism will fall away, which will erode ethic and rational differences—perhaps allowing our blandness to expose itself. On the other hand, we have no goals for technology, no rules, no coordination, so something like that might be possible. If so, it will happen in rich, defensible enclaves. And the poor will be able to view it on shared phones.

A hyperconnected global technical computer society will be a challenge, especially because we cannot process the most of connections in such a society. Spurious connections become easier to find, but may be worthless to know, for example, standing armies and stands of trees in northern lands, or margarine consumption and divorce rates in Maine. What about the internet and a search for love? The author notes that we might lose unnamed aspects of our humanity, but we have already lost many: Empathy for animals and the starving; or the ethical will to try to help make things fair and better. We do not need a hard drive to supplement or guide us.

Experiment: *Can Science Save us?*

The method of science, isolating something, then breaking it down, measuring it and comparing it, allowed a progressive knowledge of things we needed to use. Science started by using metaphors, e.g. the 'atom is a billiard ball' or the 'body is a machine.' It employed a method logic—hypothesis, testing, control, analysis, and repeat. And to characterize it, presented special properties: Quantification, neutrality, parts, focus, reduction, abstraction, experimentation, and the social construction of reality. Science used numbers to relate samenesses and forms. If it could not be repeated or compared (statistically), science had trouble presenting it.

A new branch, Quantum Mechanics, was expected to supply proof of objective reality, but it undermined it and killed it; participation in the field of study is not only not an option, it is unavoidable. Thus, science bumped into wild complexity and wicked problems, and now we have to rethink science. The comfortable logic of the scientific method—still unaccepted by a significant percentage of human beings—is frazzling. It cannot handle wicked problems (those recognized by Horst Rittel as having conflicting frames of reference).

These problems, including climate change, economic disparity, political violence, destructive agriculture, harmful algae blooms, and cancer and its costs, cause uncertainty and contradiction in the scientific approach. Science has trouble dealing with unique situations, such as a living earth

(Gaia, according to James Lovelock). Most situations we encounter are unique, 'one-off' cultural contexts. Furthermore, many are not visible, small, static, short-term, human-scale problems; they can be invisible, extended, dynamic, long-term, and large-scale.

Science could still work on simple domesticated problems, from launching a rocket to splicing chemicals, but it has difficulty with categories and disciplines. Obviously, it will continue to be effective at one level, as a special case of a new unborn paradigm. What would this paradigm be like? Should we call it 'ethnopoetics' or 'metascience'?

Beyond a binary one, it would have an expanded logic, like Chinese, Mandenka or Hopi; this would allow it to consider apparent contradictions, such as wave/not-wave. Instead of being based on a restrictive 'machine' metaphor, it would apply an organic metaphor, which could deal with patterns of interactions and connections. New words or phrases, such as 'holistic' or 'reciprocally constrained construction' (Russell D. Gray, 1988), would be needed. Experimentation would have to be enlarged for some cases, to account for the participation of the observer, or eliminated by thought experiments. The subjects would have to be kept whole in whole contexts. Despite the importance of statistics for comparing similarities, we would have to accept the unique characteristics, history and context of each subject. And, we would have to synthesize subject patterns, as well as analyze them sometimes. These beginning ideas do not exhaust such a paradigm; we are still working to describe it. Sometimes we have too many possibilities.

Science presents us with too many facts, yet we crave to have more. Philosophy presents us with too many values, but we hold too few. Technology presents us with too many things, and we do not know which we really need. There may be too many futures to choose from, but we can limit them to three possibilities: The fake ones, with amenities for some cute lodges in tame wilderness surrounded by wastes; the technical ones, with maybe a real fenced-in wilderness and some kind of technological world city; and a eutopian one, based on cultural wisdom, traditional forms and an ecological sensitivity. This last is open to new knowledge and approaches. There are ways of dealing with the earth that are not scientific or technological; they are aesthetic or ethical. They are not incompatible with a whole science. The nature we study is a feeling system, populated by intelligent beings. We trust that science can continue to undermine its own metaphors and limits, to include more of the universe.

Experiment: *Can Art Save Us?*

Art is communicative of the quality of things. Like science, it discriminates the unsuspected in the commonplace. It is not opposed to science, but more diffuse; not better than science, but more comprehensive. It accepts ontological parity, the equality of beings; aspects of the world are not negated or reduced by one another. As metaphorical knowledge, art can avail itself still of scientific references. Art can measure a whole qualitatively and mimetically, a germ or the cosmos with its imagery. Art is a tool for comprehending partially what cannot be known totally. It furnishes a view of the whole.

Artistic ideas must first be comprehended in their fleshy existence; only if they are first experienced can they be grasped, and then they cannot be transformed into ideality. These ideas, musical and literary among others, cannot be detached from the sensible appearances, as can the ideas of intelligence (and science), which are erected into a second form.

Like science, each new medium undercuts the old paradigm. Like quantum mechanics, which was the death of objective reality, which is what it was supposed to prove, art undermines its parent idea. Something new flourishes, then it is old. Then the revolution is over, and old art is part of the new environment that will require further art to see. The media become the field of the new idealistic ecosystem.

Unlike science, art uses the rare, unique, individual instance for its lesson. Science uses numbers to relate samenesses and forms. As quantum mechanics showed that participation in the field was not an option, so art requires participation for its consumers to recreate the subject themselves.

The two are different, of course, as art focuses on the unique in everything, what is different in the pattern or flesh, while science focuses on what is common, on what can be measured and compared. That is why art is metacentric and science is numerocentric.

Maybe that is the problem. They should not be centric at all. Art should be metaperipheral, science should be numeroperipheral. Does art not focus? Is there a way of not focusing? What about the periphery? Could we approach it sideways, crab-like? Can art do that? Or do we need something new?

Despite their subjects, art and science both work to expand human consciousness. They both use metaphor. They both make the invisible visible (although science claims to make the visible invisible sometimes, as with a chair and its atoms). They both use technology. But, as metaphorical knowledge, which may be prerational or metarational, art or poetry can avail itself still of scientific references.

Can art enlarge or alter the perceptions of all human beings on earth with the selection and presentation of relevant information to form an ecological consciousness? The survival of society now depends on a consciousness of the global system in its complexity and connectedness. The things required for survival—order, diversity, cooperation, and knowledge of biogeography—are things that art teaches are valuable. The spirit of humanity depends on a consciousness of its proper relation to the wild places of the earth. Art, combined with an ecological science and politics, could be adequate to deal with the creation and maintenance of good places on earth.

Archaic norms and value systems are crumbling under the images and artifacts of industrial culture. It is not just things, practices and languages that are lost; it is the stories, the skills, and the artistic creation of forms and things. Industrial culture says they are not important, as it uses its power to have bad science replace good science, as it bans undesirable topics in art or science. Art has the potential to restore or save these necessary things. But, only if the culture lets it create.

Experiment: *Are We Too Complacent?*

What happened to mad cow disease? Do you have it? How do you know? Should you worry? Over 179,000 cows had it. Many human victims have been young since the disease peaked in 1992 (with a few individual cases later). I think we should worry.

Some dead cows with mad cow disease are composted. Others are dissolved, into a soapy, tea-colored fluid. Others are burned at high temperatures (1832 degrees F). Others are rendered (ground up) into pellets, which are used in animal feed for—cows. Cows had been unknowing cannibals for decades without the threat of mad cow disease, but the practice saved consumers $0.01 per hamburger, so it must have been worth it. These processes destroy almost all infectious agents, except prions, the misshapen protein that causes mad cow disease, a fatal, incurable disease that eats holes in the brain. Almost all agents. Cooking the meat offers no protection to consumers (zero, zip, nishto, nada).

A few cases between 2003 and 2005 were dismissed by scientists and specialists as being caused by a random mutation, rather than from an outbreak in the cattle. So, it was decided that there was little risk of a large outbreak.

By 2011, mad cow disease was said to be extinct. There were no reported cases in the US. The incubation period for mad cow disease, however, is decades long, so someone infected in 2010, for example, may not have symptoms until 2020 0r 2033. After 20 years, how could you link it to one hamburger? Some time in the next 10-50 years, there could be an epidemic. It will surprise us, and we will rush to reinstate treatments.

The symptoms resemble those of Parkinson's, Alzheimer's, dementia, and neurological problems. Just the symptoms showing up in politicians, who might decide to protect the industry rather than an easy sacrifice of 50 consumers. Testing is rare and spotty. We don't know how many people have prions.

Mad cow disease is only one instance of people being poisoned and killed by introduced toxins, chemicals, gases, and radiation in the environment. The pattern is one of limited suffering, with professional laxity and public indifference. Doubtless, we will continue to be complacent until a catastrophe crushes millions.

Experiment: *No-News Crime Reporting*

In their races for ratings, radio and television corporations present the most gruesome or rancid tragedies. They often present the same ones, over and over. If it has carnage or sex, or both, the ratings will go up. Profits will go up, and that is considered good. But ethics and standards go down, and that is bad. Furthermore, the media presents the facts backwards and reverses the coverage, so the victims are ignored and the criminal is celebrated. Our culture needs to change. To celebrate or recognize the lives of the victims, not the violent perpetrator.

We need a new set of rules to present what is important, rather than what is the bloodiest or the most tortuous.

To start, do not show the criminal's face at all, unless public participation is needed to apprehend her. If possible, do not name her. No photographs of the actual incident, no psychological analysis or social history, should appear in daily media. Let academics or pop authors present it later in magazine or book form, with a significant part of the proceeds being given to the victim.

Next, do feature the victim and what she wanted or worked for. Emphasize what her life was like and what her goals were, what made her happy, and what she hoped for. Focus on the victims, on what happened and how it will affect them, and how the community can help. And highlight the detectives who solved the crime.

Finally, do not overdue it. Do not keep looking at the empty street or parking lot over and over and over, looking for blood stains. Do not replay any footage of the actual accident or murder. Do not keep inciting fear or panic about how bad things are. Put it in perspective: In the last hundred years, there have been fewer wars, fewer murders and much less violence.

If the media needs to keep analyzing the problems of unhappy people, by all means present statistics relating the crime to the weather or good luck to the stars.

Part 14: The World & Fear

Experiment: *What Should We Fear?*

Most people fear risk, pain, humiliation, torture, and death. The news presents us with many other things to fear, from rape and terrorism to accidental rest room exposures and revolution. Our culture exposes us to still more things to fear: Sex, animal bites, wilderness, cheating, automobile crashes, accidental nudity, and boredom.

Addictions can amplify some emotions, such as fear or hate, especially as they relate to the possible end of the addiction or the threat of that end. People cannot escape their addictions, whether to drugs or ladies handbags. Knowing nothing of history and expecting nothing of the future, people cannot escape the fearful isolation of the present. They join together in a melancholy herd. We fear correction and reprisal, misunderstood joy or passion.

People show a desire for new technologies, tools and products, but often fear and resist change. Resistance to change can be a normal part of a cultural process. Once we get the new technology, we fear it will break.

We fear being at a disadvantage. One advantage of a healthy economy is that businesses can make decisions about wages and working conditions without fear that these would put their businesses at a disadvantage in competition with businesses elsewhere that paid lower wages and had worse working conditions.

We fear running out of resources or food. We fear things will stop. We could build from recycled materials alone, so there is nothing to fear from stopping. Economists, however, fear that "letting things alone" will lead to stagnation, poverty, and chaos. We fear to let go of the technological vision, of "life under control."

We fear violence in our communities. C. R. Jeffrey suggested that the proper design of the built environment could reduce crime and the fear of crime, by minimizing opportunity and promoting positive behavior. But, apparently that is too expensive, and so rarely done.

And those people feel the need to protect themselves from others who are different, whether in religion, wealth, customs, or land. Fearing large destructive wars, people tolerate many small permanent wars (which are just as destructive of ecosystems and cultures). Wealthy nations fear population implosions that could destabilize the nation. They are fearful of change or exploration. We fear surprises. We cannot predict global climatic or ecosystemic catastrophes.

Nature seems distant and unknowable, so it is feared as unfathomable and uncontrollable. Nature seems contradictory and sinister, shaped by death, which we fear. We fear to understand, to be compassionate. So, we try to dominate and control nature, to overwhelm it before it can do it to us.

We fear nature, its infinity and uncontrollability. In its immense complexity, nature seems wholly other, nonhuman, ultrahuman. It seems distant. So it is feared as unfathomable and uncontrollable. We fear to understand, to be compassionate. And, fear casts out love and, with love goodness, beauty, truth, and intelligence. Until all that remains is fear of other beings and the unknown; fear of the smiling science and technology that takes away more than it gives; fear of our fellow human beings, who are trying to regain what was taken.

But, love can cast out fear. In the Upanishads it is written that "Who sees all beings in his own Self, and his own Self in all beings, loses all fear." As fears and unconscious motives are understood, the awareness of all feelings intensifies. Feelings that are dualistic at one level—fear and courage, pride and humility—are combined at a higher level. Unconditional love blends many feelings that cannot be understood at an intellectual level. If we could just overcome that one great fear, how to love without conditions.

Experiment: *Will Industrial Civilization Collapse?*

The real question is whether any civilization or culture that human groups have formed is sustainable for very long. We know that foraging groups (hunters & gatherers), and some herding, horticultural and agrarian societies are sustainable, mostly because of their acceptance of size and limitations. But, we do not know if this is or will be true for modern civilization, especially an urban, industrial civilization.

Many cultures or civilizations have collapsed and disappeared, often without being remembered or documented. The known collapses range from the Old Kingdom of Egypt (2181 BCE) and the Harappan in the Indus Valley (1750 BCE) to Rome (476 CE) and the Khmer in Cambodia (1431 CE). Recent collapses include the Kachin of Burma (1950 CE) and the Ik in Uganda (1970 CE).

Collapse is the rapid, significant loss of an established level of complexity after some catastrophic event. Collapse can occur as a result of a single reason or combinations of circumstances, from invasion to social dysfunction, economic inefficiency, and political dysfunction. An example of social dysfunction is class conflict. In the Mayan lowlands, class conflict was complicated by militarism, overtaxation, and land degradation. These contributing factors may have been solved, except that three long droughts weakened the Mayan cities further until they collapsed.

Joseph Tainter considered that complex systems were profoundly maladaptive, since political responses to stress became less flexible. Complex societies put harsher demands on local environments. Political regimes set production demands based on needs rather than on ecosystem productivity.

Smaller urban civilizations have collapsed, sometimes for reasons such as weakened support for excessive overhead, transfer of wealth to private individuals from public, and the breakdown of civil society—problems that are identified in our urban society now. In some cases, such as Maya, environmental degradation did play a role in collapse of the civilization, either as a cause or effect.

When earlier civilizations collapsed, however, they were isolated from groups by natural barriers as well as by looser links of trade. The drivers of the collapse were local or maybe regional. In general, groups were self-sufficient. When Maya civilization collapsed, at least the urban political structures, the people were able to return to simpler forms of foraging or agriculture.

In an interconnected global civilization, massive ecological or social failure in one region might threaten the stability of the entire global system, if people cannot go back to a state of relative isolation and self-reliance. Furthermore, the role of chance or contingent events becomes larger, as droughts, shortages or diseases affect different groups or areas. Geoffrey Chew suggests that the collapse of civilizations is not really a problem for learning, but it is a problem of not learning, as when cultures choose not to learn from past mistakes. Of course, like death, collapse is hard to learn for the dying or collapsing entity—in fact the lesson cannot be learned if the collapse is complete.

In order to avoid collapse, a culture has to learn from the past. But, nothing in our past has prepared us for the number and strength of global connections. During any collapse, many cultures have been able to adapt and evolve by expanding into other areas, but this is no longer possible in a completely occupied, territorialized, owned planet.

Every civilization faces a spectrum of risks, which may push the civilization to find new sources of land, food or energy. The need to solve such risks pushes the system further from its original state and can accelerate the operation of the system. Acceleration makes decision-making, especially weighing alternatives, much more difficult. The conjunction of long-term risks can lead to a management or governmental crisis.

We have never planned or designed our civilization to respond to impacts and the dynamic changes of the entire planet. We never designed civilization to respond to complex challenges or to slow or invisible catastrophes. Thus, we are not responding to many kinds of catastrophes, and when we do, we only shore up a quick fix, regardless of the regularity of the catastrophe. Sometimes we may not recognize that the catastrophe is an effect of our designs and activities. Sometimes when we do recognize that, we think that we can use the same mindset and approach to correct the problems later, when it might be more economically or politically feasible, yet it rarely is.

Furthermore, we are engaging in several large-scale experiments, which can conceivably impact large areas of, or possibly, the entire planet. These experiments are not just our civilization, but also our way of growing food, traveling, and converting landscapes. They are experiments on the change in the makeup of the planet, because we select so many different kinds of organisms that we need and we transform so many ecosystems. Civilization and cities are large parts of that transformation, now. Our investment might work, or it might collapse.

Experiment: *Are There Patterns of Collapse?*

There are many types and ranges of collapse. In some cases, the population collapses, but there is not much loss of cultural complexity. Ireland, for instance, lost half of its population, during and after the potato famine, because of its reliance on that one crop; due to English rule, the government continued. The population stabilized at a new lower level, under three million. After 1960, with the development of industry and then membership in the European Union, the population has been increasing to six million, still less than the eight million in the 1800s. And there is less reliance on one or two crops.

In other cases, there has been a tremendous population loss, along with a political crash and complete loss of organization. The Maya collapse led to the complete abandonment of the cities and the urban way of life. The population dropped dramatically. After people dispersed to return to older styles of living, the local ecosystems regenerated, without the pressure of use or overuse, and recovered.

In still other cases, local city states may collapse for a generation or two, and then be rebuilt in place, as the local ecosystems have experienced some level of renewal, perhaps at a lower level of complexity themselves, where the forests may be gone, but the grasslands have recovered. Mesopotamian civilizations recovered numerous times, as new states were able to benefit from the natural process of desalinization of soils in time.

The ideas produced by a culture may spread and reproduce, as people move to other places. Lower levels may survive with many of the ideas of the larger complex. So, the Roman Empire collapsed, but not parts of Italy or France. Germans returned to chiefdoms.

Industrial civilization has the potential for collapse. Elman Service uses a biological analogy, where social organizations are modeled as plant or animal species that are initially successful because they are adapted to a niche, but later become overadapted and less flexible. Collapse is part of a natural cycle, according to Service, who believed that complex systems were profoundly maladaptive, since their responses to stress became less flexible with time. This argument does not consider invasive species or the maturity of the system.

In a systems model, collapse is part of stochastic process, which implies that civilizations will die, but not necessarily within a definite time frame or for specific reasons, such as overconnection between trading groups and social classes. This could be a problem with modern globalization, in the 1990s, as well as with simpler regional connections, such as in the 1300s between Asia and Europe. In the general system model, complex systems are hierarchically composed of many stable lower and intermediate orders, strongly connected horizontally, but less so vertically. The problem may be more one of scale.

Human cultures tend to fill all available space, carrying capacity space, even though some of the spaces are occupied by other species or other cultures. This makes them prone to crash after sudden changes. They have adapted to marginal environments with behavior and technology, such as that for water storage and grain storage, to buffer themselves against the known changes of the environment, but when an unknown change happens, such as a nine-year drought, they fall apart. Few people have the luxury of moving to a virgin area.

Any one strategy, such as reducing the human populations—without making changes in distribution, pollution (related to climate change) flexibility, and many other factors—might not let us avoid collapse, but merely postpone it. Reducing the population would relieve stress of ecosystems and reduce the destruction of species and systems. But, if we continued growing economically and connecting tightly globally, then the catastrophe of a single key resource, such as cheap oil, could set off cascading pulses of contraction and collapse.

If we keep enforcing historical inequities while relying on a lifeboat approach to saving civilization, that could set off havoc that would lead to collapse. Reducing and refitting our massive appropriation of ecosystem productivities and services would allow those systems to recover and diversify again. But, if we continue to rely on externalities for free services, without restoring and setting aside a majority of the planet for wild systems, the support system may break down and trigger ecological collapses. By becoming less complex and connected, by deciding on a satisfactory level of sophisticated culture, a mature civilization might have the flexibility to recover from a collapse if it was unavoidable.

Lewis Mumford noted that each civilization begins with a living urban core, but ends in a common graveyard of bones and broken pottery. Every complex culture so far has failed, although some like the Romans or Chinese, have been able to renew and rebuild subsequent cultures. Collapse has been history, but it is not necessarily fate. Cultures do not choose to collapse, despite what Jared Diamond says; they choose to continue successful behavior in inappropriate circumstances, in spite of the unsustainable costs—the Greeks called this the operation of tragedy.

Experiment: *Is There an Alternative to Collapse?*

We need to recognize that our catastrophic situation requires immediate
emergency actions on a global scale, from expanding personal
consciousness to reforming the very character of human civilization and
its coevolution with wild nature. First, an international organization such
as the United Nations has to assume control of the planetary commons
for limited utilization as well as protection. It would manage the commons
in the interests of all living communities. By charging usage, extraction
or loss fees for water, fossil carbon and other resources, it would become
self-supporting. By international agreement, it would have the largest and
only standing army, with some large-scale nonnuclear weapons—nations
and communities maintaining police forces for public safety and control.
It would allow all nations—even minor landless ones—to join with
equal votes. It would protect traditional and modern cultures in a loose
framework, encouraging cultural health related to environmental health,
stressing economic equity through a variety of measures, and trying to
direct the emerging global culture. Through this organization, we could
consider incorporating the planet as an interested party.

This international organization, as well as nations, communities
and individuals, would apply an ecological perspective to knowledge and
processes and, to understand how communities and ecosystems work. Every
person and group needs to understand how challenging events can become
problems, especially water flows on land and atmospheric change. They,
we, need to face the global problems of a chaotic changing planet and to
comprehend the limits and weaknesses of cultures, especially bad images
and incomplete cosmologies. Creating an ecological cosmology, with new
metaphors and logic, could be an important step.

We need to emphasize science, especially addressing systems and
synthetic thought. We need to end the thoughtless, large-scale experiments
on ourselves, living communities and the planet, and imagine real thought
experiments. We need to reduce our massive interference with ecosystems
and landscapes, especially through ecosystem conversion. We need to reduce and recollect many kinds of pollution, from plastic nurdles to carbon
dioxide gas. We need to create surveying programs to inventory places and
the planet, then monitoring programs to keep informed as well as to reconnect many plant and animal patterns.

Human populations should be related to ecosystem productivity, as
well as to rarity and diversity, even if this means a large reduction in numbers. Failed cities should be removed or refitted. New cities should be built
with psychological and ecological prospects, perhaps as arcologies. Technology could be reduced and integrated to fit needs and limits, instead of being
automatic. Many predominantly technological problems, such as carbon
production, fossil fuel extraction, nitrogen pumping, phosphorus loss,

and pollution, can be solved through a combination of conservation and balancing. New smaller scale forms of technology can be designed using precautionary principles. Industries could be reformed as limited artificial ecosystems. Appropriate technology could be directed and certified.

We need to learn to be healthy individuals in living communities, to reduce antibiotic use and unnecessary luxuries. In fact, we need to place health in the context of ecosystems and global cycles. We need to learn to adjust to our own psychological and social limits. This means addressing problems with patterns of growth, inequality, poverty, dominance, and slavery. We need to link economics to ecologies, especially at the global level. We need to redefine the nature and limits of corporations, especially with new goals and community responsibilities. We have to consider the problems of intolerance, conflict and war within a frame of ecological ethics, and to reconsider the goals and responsibilities of individual nations and a global community.

We can design local systems and regional patterns; we can design the entire planet as a whole. We can integrate traditional adaptive patterns, from agriculture to technology and cities, with natural processes. We can consider our civilization and planet as part of the solar system and local space. We can strengthen religions with common understanding to bind people to their places and planet. We can make politics conscious and fair, with traditional goals and limits. We can make global designs, mostly to restrain humanity and restore balances of systems, but also to coevolve with the development of natural systems. We can create a planetary framework with clearly delineated functions at each level from personal to planet, based on an integrated approach of koinomics. And, we can advertise this to everyone so they can participate. Then we, they, all of us, can say what else should be done.

Perhaps our efforts will be as naïve as continuous war, or as utopian as ecosystem destruction, but we have to make the attempt, and to continue without pause or stop. Perhaps the ideas of global commons and small nations are as unrealistic as an infinitely growing economy or as 'warring for peace,' but we have to try new patterns. We have to promote the appreciation of places, to awaken the delicacies and qualities of designs, to plan frameworks for development, and to allow everyone to participate in creating good, personal and social designs. Each single step is easy. Each challenge can lead to an adventure. Each change will have some effect. Each vision could inspire more efforts. Collapse of some kind may be unavoidable, but through appropriate designs and actions, we may treat collapse as a temporarily negative event in the path to a eutopian form.

Experiment: *Could We Remake Civilization?*

We have never planned or designed our civilization to respond to impacts and the dynamic changes of the entire planet. We never designed civilization to respond to complex challenges or to slow or invisible catastrophes. Thus, we are not responding to many kinds of catastrophes, and when we do, we only shore up a quick fix, regardless of the regularity of the catastrophe. Sometimes we may not recognize that it is an effect of our own designs and activities.

Many problems are not global, regional or ubiquitous, and can be solved locally. Change can result from accident or mutation. But, conscious invention can produce new structures with new capabilities. If a global civilization is to form and to survive, it must develop into a completely new type of social system. A blind, consumer society cannot be transformed into a conserver society without constructive change.

Because modern civilization is dynamic and developing, it has the potential to change itself to be more stable and less intrusive. Individuals, groups, and nations have the ability to design workable social and living systems. We do not require more data or knowledge before we start redesigning systems. We have enough for a practical approach.

This strategy needs to 'dedestruct' the negative impacts of civiliza-tion. We could design some self-regulating feature to lessen environ-mental stresses when they become apparent. Another approach is to 'dewesternize' civilization; other cultural paths are viable. We also need to 'deglobalize' those things that should not be globalized, such as traditional cultures. We need that to 'dehomogenize' the world. We tend to be drifting toward a more homogeneous world since our weak global culture tends to be bland and generic. Yet, if it wipes out all other cultures, there will be no rivals or successors. Traditional cultures are not failed attempts to be industrial; they are successful manifestations of experience in specific places. They allow us to be aware of our place in that area of the planet.

One solution is that everyone has to become a designer, and to par-ticipate in building a culture, similar to a foraging society, where we made everything ourselves. The future of civilization is our common design proj-ect, as Morris Berman states. He suggested that designers have more power than they realize, especially in critical situations—and ours is critical. One grave threat to our future is overconsumption, which is fueled by shallow design. We must leave a better legacy by using our best ideas immediately, and not rely on copying our chromosomes for later. We have made many major errors in the unplanned growth of civilization, from our agricultur-al conversion of ecosystems to the production of plastics and synthetics. Human civilization cannot afford major mistakes. We have to try balancing civilization as an exciting phenomenon in a wild planet.

Experiment: *30 Day Countdown—When To Panic*

I remember an exercise from a math class in high school, to show us how strange and unintuitive doubling times were. If a water lily growing in a pond doubles everyday, and it takes 30 days to cover the surface of the pond completely, on what day will the pond be half covered? The answer is day 29. So, if you plan to start removing the water lilies, it might be better to start about day 5. I think the other juvies remembered the example with gold coins better, but they taught the same lesson, and that is how quickly things that double, whether toxins, populations or coins, can become problems (or wealth).

I wonder sometimes if our catastrophes and emergencies are not like this somehow. So here is a crude thought experiment to try to find a tipping point to save humanity and the planet, using a 30-day period. There are bad weather, damaging storms, noticeable extinctions, and worrisome losses, that seem to be doubling before Day 1. That is just prelude.

For the first two weeks of this project, we argue in person, in blogs, and at official meetings. Then, we get down to business.

Days 1-14: Converse, warn, predict, shout, ask for solutions.
Day 15: Notice absence of animals, increase in pests.
Day 16: People are suffering from stress.
Day 17: The content of the atmosphere shifts. Heat rises rapidly.
Day 18: Weather patterns change unpredictably (chaotically).
Day 19: Scientists document 75 new extinctions.
Day 20: Unseen elemental cycles stop working sporadically.
Day 21: Ocean acidity increases; greenhouse elements outgas.
Day 22: Weather more intense; storm surges drown 17 cities.
Day 23: Fewer birds are seen; strangely, fewer butterflies, also.
Day 24: Wild plants and trees die faster from a bad fungus.
Day 25: Agricultural systems start to collapse, as crops fail.
Day 26: Military conflicts begin for strategic minerals.
Day 27: Rains become wildly variable, more droughts.
Day 28: Domestic forests start to turn brown. Insects disappear.
Day 29: Military conflicts for food. Less oxygen, hard breathing.
Day 30: Global biocycles stop. Most humans die. Life retreats.

Where was the tipping point? One seems to be about Day 20. Another is Day 23, when living webs are observed collapsing. Once system feedbacks change from negative (which is stabilizing) to positive (which is destabilizing), corrective change is almost impossible. But, the effective tipping point was probably Day -243. We just could not see it then.

Experiment: *We Reach an Optimum Population. Then What?*
We have calculated an optimum global human population and discussed
how to reach it, but 'Then What?' as Garrett Hardin asks.

So, assume we reach an optimum number, using resources at
replenishable rates or below, and are in equilibrium. Is there one ideal level
or a sliding scale depending on change? We would have prosperity for a
while, but then face new challenges from a dynamic planet. Mad growth,
or going beyond initial growth past maturity, is not necessary, but maturity
means continual development and learning. Our knowledge and ignorance
will not stay still or become a constant utopian boredom and stasis.

The population could shrink to an optimum, one of many possible,
then adjust to conditions after further experimentation. The economy
would have to adjust. The initial adjustment should be easy, since we have a
temporary surplus of energy and manpower to disassemble and reassemble
cities and shared economies. Our creativity may limit our plunge or our
courage may limit it.

Population growth has brought prosperity for many at the beginning
of the 'chain letter.' Resources are overused and discarded. Values have
to change from forced consumption to treasuring things. Prosperity has to
mean time for families, for play, for learning and interacting with the wild,
and for continuous inspiration and expression.

Agricultural output is shrinking and may shrink more with a switch to
organic farming. Output from the oceans is unsustainable and falling; it will
fall more as we let fish populations recover, for as long as it takes. Or, if the
ocean crashes, the output falls to zero. Population will follow, especially as it
ages. Perhaps the aged will have to keep working and being creative longer.
Population has to fall to avoid famines; famines will reduce the populations
with much more suffering to the same levels or below. That could happen
even if we never attempt to reach an optimum, to try to reach a balance
with other species on a wild planet.

Figure 196. Designing
Civilization in Thomas
More's *Utopia.*

Experiment: *Could Humanity Destroy The Earth?*

Could we destroy the earth? *Yes*. If we wanted to, of course.

We express our fears about our damage to the environment as a planetary catastrophe at a global scale. But, this destruction, diversity loss and extinction are not dangers in planetary time. Both pro-environment and pro-development lobbies use inappropriate time scales to worry about diversity and extinctions.

To be sure, these crises are real and indeed frightening, but they are human crises, species specific and community-oriented. We are losing the species and communities that inspire us to love them. Our moral questions have to be answered in the small span of human civilization, of human biological and social time. If we are mature enough to ask them and to respond appropriately.

How much longer should the earth last? 5 billion years? 2? Despite the pessimistic predictions of some scientists, that the earth is already old and senile, and will not even last another 1 billion years, as a planet the earth should last until the solar expansion destroys it, in about 4–5 billion years. As a living planet, the age is harder to predict. In the oceans, or in wet rocks under continents, life as anaerobic bacteria and archaea might survive for another four billion years—even though the bacteria may not be able to support an atmosphere with biogeochemical cycles, or a web of more complex forms. More complex forms of life may be vulnerable to sudden events or vortices quite soon.

How long will humanity last? Ten million years? A billion? We likely will not last a million years, much less a billion. But, within that time, what we do is important. It is important to save communities of species, even if they will be extinct in 3 million years. It is important to respect and permit the process of evolution, the whole dynamic renewal of life in the planet.

Human consciousness and the ability to consider the long-term impacts of our behaviors in a civilized context, is important. Human cultural change does speed up feedback to biological and physical evolution. Culture is transforming the surface of the planet. But, why not transform it wisely to protect all the things that support good human life? Perhaps human culture could double our human persistence to 5 million years? Or six. That would be amazing and worthwhile, but it cannot be done with rapid ocean rise and atmospheric pollution. We are not immune to the calamities that we cause. And, it may not be possible to survive as long in artificial environments, such as cities or spaceships.

With our toxins and nuclear devices, we could seriously alter the ecosystems and cycles that support complex life. By accident, or through ignorance. But, I pray we would never want to destroy everything in a fit of rage or intentionally through many greedy little steps.

Experiment: *Are We Doomed? Yes!*

I know, it's alarmist. Defeatist. Depressing. Negative. But we *are doomed*. It cannot be denied. Look at the evidence.

People are contrary. They don't believe in evidence. They cannot agree on the same things. And, they refuse to cooperate, or to admit someone else might know something and be right, if they didn't think of it. Well, that's just the mob.

Now, it's true that a mob in a limited horizon can choose things and directions wisely. But, apparently a mob does not look beyond the immediate interest—certainly not ahead for 7 generations, and certainly not at the entire human collection of mobs. Those other mobs are too different to be useful or wise.

Economic growth cannot save anything. It is petty and short-term. Looking at profit without a worry. Creating wants and filling them with junk. Bankers and financiers do not even bother trying to balance a budget that has 276% missing from internal thefts and incompetent investments in hidden packages.

And politicians. Please. They would polarize a polar bear if it got them elected, so they would not have to take their fingers out of the cookie jar (or their underwear). Someday something will happen in a two-year stretch, and they will be ready. At least ready to start planning. But most things happen in days, weeks, months, decades, centuries, and millennia (or longer beyond the human scale), not a perfect 2-year frame.

The technologists cannot help. They say technology will save us at the last minute with infinite energy, but everything is getting torn apart now by excessive energy. Torn, and dirtied, and bankrupted. Maybe nanobots can put it together, but I suspect they'll just make more nanobot luxuries and Npads or something.

Scientists. They know so much, but they do not put values on it. They have facts and figures, but no idea about how to apply them. It might disrupt the holiday season or the new TV schedule. Or their vacation schedule. Or their comfortable relation with their employer. Something important. Something that will prove their scientific neutrality.

We don't stand a chance to change. Look at us now. We rent movies and buy cars, even some that get 40 miles per gallon. How sad. We build bigger houses on smaller lots and fill them with timesaving appliances until we have time to look around and see—what?

The cliff approaching? We can't change our diet, much less the momentum of the industrial vacuum cleaner. We can't stop watching TV and worshipping actors, much less change our behavior (and our images and goals) to survive another 20 years. We can't even have civil dialogues, much less work together to repair the damages done in the name of progress.

Like a chameleon, we can change our colors, but not our shape, not our behavior, so buy that 90-inch TV and you can watch the awful collapse and suffer the sudden realization that with just a little easy—well no, a lot of difficult change earlier—we could have lived longer and been healthier and happier. And, civilization and the planet would have been more predictable and comfortable

We're just *doomed*.

Figure 199. Doomed & finally noticing the cliff (Credit: Google).

Part 15: Globalizing Everything

Experiment: *Is there a Global Culture?*

Human societies have tended to grow larger over time. With the change in scale of populations, other changes have occurred in the structure of settlements, fields, governments, and religion. These changes may foreshadow some kind of global culture or government.

Some cultures grow, and in growing, decide that growth is a good thing to have without limits. Many lucky accidents, such as the rediscovery of the Americas, gave some cultures an impetus to keep growing. Other developments, such as science and industry, followed the conditions created by plagues and environmental restraints. The Spanish became a megaculture after benefiting from their exploration and exploitation of the Americas and western Pacific. The English became a megaculture after establishing an empire from North America and the Caribbean to Africa and Asia. North America became a megaculture in the Twentieth-Century. China is claiming the title this year. Each megaculture was able to dominate part of the planet with its influence and to see some of its products or rules become ubiquitous.

It isn't that traditional cultures are excess baggage, it's just that they need to have a global dimension now. Maybe this is a domestication of cultures. Maybe self-domestication is adaptation to a global niche. Global capitalism undermined many traditional cultures by offering consumerism in the place of traditional cultural guides for behavior. Social roles seemed irrelevant by comparison, if the good life could be bought without effort. Yet, it did not seem to work in Europe and the US. Instead of being free from economic want to develop their potential as creative human beings, people became trapped in a consumer cycle. Self-actualization was postponed for self-gratification. Democracy seemed to be good for balancing a middle class in some cultures, but it ignored other cultures and economies. By virtue of its demomass and political system, China is becoming a megaculture. Chinese products are dominating the economies of other nations.

Trade and exchanges by various cultures extend to distant lands. Over time, a global system starts to develop, with newer connections and technologies, which draw the cultures and civilizations, of all kinds and ages, into a tighter pattern. Some global civilization may result, but what effect will it have on the earlier patterns? So much is lost, of ways of being and acting in a more natural, less technological environment. The world has acquired a rough, global economic structure.

Culture stretches vertically to include the physical, economic and political. It stretches horizontally to include society as a whole. In fact, culture is concerned with all things and beings. It is organic, like Aristotle considered a work of art, and whole. So everything in it is interrelated to some degree. Many relationships are encompassed by the holistic perspective of culture: The relation of people to themselves, to each other, to objects they create, and to their natural environment, and to their cultural environment. These bear on psychological well-being, social bonds, material legacy, and on the association with other forms of life.

Yet, we do not seem to have a unity of culture because every culture requires a local place, and specific particulars. Some, including Modernism, Marxism and Liberalism, see the trend to a world culture of poets and scientists. Others, from Traditionalism to Conservatism, see a global culture bound by economics.

Global exchange, however, occurs at a different scale and a much faster tempo, so that cultural elements often sit beside one another without any kind of meaningful integration. In these cases, the scale of exchange may overwhelm the formal possibility of a holistic culture with an assortment of materials that are personally chosen. Some items, such as blue jeans, give a superficial uniformity to all cultures. Global exchange seems to create a superficially tight integration of economics and cultures.

In conclusion, there is no meaningful global culture that all of humanity participates in. It might be possible, but it has not condensed yet.

Experiment: *Is a Hyperadaptive Global Culture Forming?*
Is there such a thing as a global culture? Small cultures are being drawn into national cultures. The disappearance of peripheral cultures might be a calamity, but, modernity, in the predominant versions of liberalism and Marxism, sees the goal of history in a universal world culture. Alas, one wonders if a global culture would be worse than the national cultures that are so irreverent with their peripheral cultures and environments. In destroying peripheral cultures, there might not be any culture left to be the basis of a global culture.

The 'McDonaldization' of the world is about economic dominance, the homogenization of trade networks and consumer rituals. This may undermine the identity of other cultures, but does not make a global culture, the idea of which has become a consumer object as much as burgers, fries and coke. Other local phenomena, such as Hollywood movies or Chinese slippers, that extend around the planet are local things that have been globalized. Cosmopolitan travel has the same flavor. Some cultures have more power because of these competitive advantages.

Although modern communications technology could work the same, it also may allow cultural differences and complexity to remain. The center seems to yield to the periphery, as other cultures all have hotel rooms, burgers, slippers, and things.

People can react by emphasizing the uniqueness of their ethnicity. No real globalism can exist until cultures and nations create a framework for the use of and protection of the cultural capital of the planet, and until cultures enter into dialogues about it.

There is an upside to a global culture, which could emerge from the interactions of local cultures, and would allow for more rapid exchange of ideas and things between individual cultures. It would provide the opportunity for trade in special goods, which have the potential to benefit every culture. It would provide paths for communication that might stimulate cultures to learn from one another. A global culture could present a global morality, built on universal human tendencies, that would not be prescriptive of our private and public behaviors, but it would be proscriptive of damaging behaviors from murder and obsessive greed to the wanton destruction of ecosystems. Global functions and problems, such as atmospheric warming, would be easier to address. New global economic or political structures would also be easier to form.

But, there is a downside. Groups like the Pygmies have specialized to fit the requirements of their environment, successfully. This makes it difficult for them to adopt other cultural arrangements. The local and the particular are required parts of culture, so global culture has a contradiction. We may not need a unity of all people any more than of all wolves or fleas. Is it wise to have a single world? A single market with a single control? It might create a suite of problems.

Hyperadaptivity is a serious condition that allows humans to adapt to poverty, bad diets, crowding, stress, suffering, and immense natural loss. Of course, we are unconscious of many of these problems. The homogeneity of forms promoted by a global communications and advertising campaigns leads to loss of diversity. A severely limited number of options leads to a lack of flexibility and stagnation. Then, we have the hyperpersistence of error, with stupidity and violence, which forms a positive feedback loop. At the same time, interactions are accelerating. That makes the baggage of tradition more difficult to carry and use. We think the past is over and has no hold on us, but then we lose all the tools and directions for living in place. A global culture seems a bad idea for the near future.

Figure 202. Read first, then build? (Credit: KornFerryInstitute)

Experiment: *Is Globalization Destroying Cultures?*
To increase the economic wealth of global economics, we have created
a 'global marketplace' in a 'global village.' We have tied together people
with millions of televisions, hundreds of millions of telephones and billions
of radios. More and more people eat the same foods, wear the same style
clothing, and read, watch, and listen to the same entertainment. People are
pressured to give up their ethnic identity and kinship for the 'global unity'
of humanity. This global culture suffocates local cultures, unique dialects
and ways of life.

Progress is erasing archaic cultures. There have been great cultural
transformations over the past 10–20 thousand years. Archaic or primary
cultures regard the relationship of human beings to nature as one of
kinship; all neighboring beings fall within moral consideration. These
cultures observe the synchronicity between their bodies and nature, and
understand their culture through mythical explanations. They employ
hunting, gathering, and shifting agriculture.

Secondary cultures analyze and deduce the operations of nature;
rituals become more stylized. Cultural innovations permit larger human
populations; ecological limits are raised by agriculture, although they are
not eliminated. Moral consideration is reserved for human beings and
sometimes other conscious beings.

Tertiary cultures (in fact, the real meaning of a third world,
twice removed from nature), are based on mechanical images that
objectify nature. Drastic changes in the production of goods forces
other psychological and social changes; human relationships are based
on economic allegiances instead of kinship and exist in societies instead
of communities. Money becomes a symbolic representation for the
value of labor and land, which are considered mere commodities.
Social stratification and the specialization of labor become fundamental
characteristics. Social and economic orders are rearranged during the
process of urbanization. Moral order, for example, becomes subordinate to
technical order.

Population pressures, resource shortages, and manufacturing "side-
effects" cause instability in many societies; militarism, intolerance, crimes,
and health problems are symptoms of that instability. Confusion and
misinformation contribute further to the destruction of cultures. The
instability of cultures, as well as stress, insecurity, and insufficient diets,
results in psychological problems for people. Individual powerlessness and
disillusion provokes the further disintegration of cultures.

The images and diseases of secondary and tertiary cultures had
immense repercussions on primary cultures. American cultures had no
resistance to diseases bred in European farms and cooked in cities. Many
cultures could not compete with more aggressive groups. Many primary

cultures have lost 60 to 99% of their populations. The Tasmanians, for instance, lost over 98% of their population; that much stress on a culture usually results in extinction, as happened to the Ona and Yahgan in Tierra del Fuego. Over 30% of the groups in Brazil were gone by 1957.

Some cultures are simply wiped out. The Herero people in southwest Africa were exterminated as a culture by German forces. The Yanomami and others in Brazil are facing threats from prospectors and ranchers, now. Other cultures subside or intermarry out of existence. The Birale people in southwest Ethiopia have only 89 remaining members—and only 19 of them speak the tribal language, Ongota.

Industrial culture is wrongly considered to be the evolutionary successor to primary cultures and is displacing them rapidly. Scholars once plotted an evolutionary trend of cultural types, from primitive through historic, modernizing, and modern; they speculated that later developments were more adaptive than earlier ones and should replace them. It was assumed that the modern view culminated from earlier ages; thus, the 'superior' modern cultures were justified in exploiting or removing 'primitive' cultures.

There is no evolutionary trend of cultural types from the primitive to modern. Later developments have not proven to be more adaptive than earlier ones; nor do they necessarily replace them. Ethnic groups are not anachronistic stages that point to Switzerland or Japan; they are equally valid ways of living. Any culture is only one of many possibilities, one way of living in a unique place—there is no single correct way.

Industrial culture, depending on its expanding market system, is becoming unstable—worse, it is attempting to become a global system at the same time. We know that cultures can destroy their ecological basis, but we do not know how to extend their existence or expand one to a global scale.

Experiment: *Is a Global Culture Even Possible?*
Culture is a symbolic system of shared beliefs, values, customs, behaviors, and artifacts that emerges as a unique, coherent whole pattern that orders the experiences and meanings of its members and allows it to adapt to changing environments; it is transmitted to succeeding generations through learning behavior and language, so that culture and the environment constrain and construct each other over time as ecological, social and historical processes.

Cultures appear to have universal characteristics because humans have similar physical and mental requirements. All people live in groups, have some form of shelter, and have an incest taboo for instance. Even though a process may be universal, the implementation and symbolization of it may be quite different and unique. Some scholars have argued that these universals could allow a global human culture. That may be true, although the relationship of a local culture to a global culture may be problematic.

Over time, regional systems start to develop, with newer connections and technologies, which draw the cultures and civilizations into a tighter pattern. Some global civilization may result, but it may not cohere. So much has been and could be lost, especially ways of being and acting in a more natural, less technological environment. The world is acquiring a global economic structure through increased trade.

A global political system is emerging, as there is increased vertical differentiation, evolving from nations and regions, which leads to differentiation into political groups and economic interests. However, the world does not function very well politically at a global level. It functions in the absence of a common culture or language. There is no world law, although there is a set of rules regulating international behavior, which are generally observed and understood. There is a small homogeneous subculture, which belongs to the rich elites of every nation, which integrates cultures to some extent. The modern world has become a very interactive system for most classes, especially with computer and communication technologies, which can lead to totally integrated mass communication and extreme compression.

It is said that the emerging global system has no center. That is a good thing. But, it is expanding without limits and that is a bad thing. It could cause the disintegration of natural systems that are interlinked with our economic exploitation. Perhaps this emergence can be linked with a major cultural revolution of consciousness. Perhaps consciousness can create a global culture. We think that instant communication and creative invention are enough to make a global culture of everyone on the Internet. But, culture is not something thrown together—it has to mix and simmer for decades or generations, and sort out what has been useful. We are not at that stage yet, and it has never been tried at a regional scale, much less a global scale. Maybe the domestication of cultures is a practical adaptation to a global niche.

Global exchange could create a tight integration of economics and cultures. But, there is no direct, immediate feedback to counter or correct overshoot. The wastes become global. There is feedback, but it is delayed for a long time by the size, flexibility and redundancy of the system. There will always be delays, since people can react instantaneously. These things might inhibit a global culture, unless we can anticipate them and make them part of that culture.

Experiment: *New Taxes To Replace Old*

Taxes could be ways of internalizing the costs of the community and some artificial constructs, such as corporations. This term 'tax' in its most extended sense includes all contributions imposed by the government upon individuals for the services of the state, by whatever name taxes are called, whether it is tribute, tithe, talliage, impost, duty, gabel, custom, subsidy, aid, supply, excise, or other name. Many traditional taxes require complex schemes to avoid cascading taxes or the regressive unfairness of some taxes. Many taxes were *ad hoc* additions to a tax code to add income from new or overlooked profitable activities. These include State of Being, Event, Sales (and Value-Added Tax), and Income Taxes (the Sixteenth Amendment to the U.S. Constitution empowered Congress to tax "incomes, from whatever source derived, without apportionment among the several States").

The combination of these taxes often makes them regressive. In terms of fairness, sales taxes are generally regressive, that is, poorer people tend to pay a greater percentage of their income on sales taxes than richer people, because they generally tend to spend a far higher percentage of their income for food, clothing, shelter, and medical care. In some locations, items such as food, clothing, or prescription drugs are exempt from sales taxes, ostensibly to alleviate the burden on the poor. Some of these exemptions, such as exemptions for clothing or prescription drugs, may actually have the opposite effect. Some of these taxes discourage the efforts that they tax. For instance, a tax on value added by labor to commodities could discourage labor itself. Most of these taxes need to be eliminated.

New Kinds of Taxes. When Arthur C. Pigou introduced the concept of 'taxing bads' in the early 1900s, economists and politicians discussed the idea, but its use was limited. Environmental taxes are functionally nonexistent in the current tax code of the United States. A few such taxes have been proposed, most recently a BTU energy tax during the Clinton presidency, but the defeat seems to have discouraged any kind of environmental tax since then.

Taxing 'bads' is essentially a 'tax shift.' Taxes are reduced on things that should be to be encouraged, such as work or savings. The loss of revenue is compensated by taxes on 'bads,' that is, on things to be discouraged, such as pollution and waste. A tax shift could help mitigate the impact of pollution charges on some businesses; it would remove taxes that discriminate against low-income families. A tax shift might also improve general fiscal health, as well as over-all environmental health.

Rather than use old labels, which were the result of *ad hoc* additions to many tax codes, it might be simpler to present them as what they are taxing, that is what kinds of income we want and what kinds of behavior we want to encourage. These kinds of new taxes can be put under four categories: Use, Loss, Adjustment (or the misplacement of resources, such as pollution), and Distribution taxes.

Use Taxes. A use tax is similar to many consumption taxes or a few severance taxes. These taxes take a percentage for a service for any resource. A use tax would have the effect of limiting the use of nonrenewable resources, such as coal or oil, as well as the use of slowly renewable resources, such as forests. The rate of the tax would be related to the scale of the economy, as well as to the carrying capacity of the ecological support system. It would be useful in monitoring vital resources, such as air, water, land (ecosystem), elements (sulfur or phosphorus), and species. Tax collectors would monitor points of entry, from wellheads to forests, to ensure that the tax would be fair, and that it would be paid. The tax would be easier to collect and harder to avoid than income taxes. It could be, and probably would be, included in the cost of any commodity that used the resource.

Loss Taxes. There would also be major loss and adjustment taxes. Loss taxes would be applied to land conversion, fossil fuels, geothermal energy, slow renewables (forests, fish populations), and fast renewables (hydropower, wind power).

Adjustment taxes: Sin (alcohol, cigarette, drug), pollution, personal pollution, industrial, and agriculture (pesticides/fertilizer). Also to the sale of heritage items, financial speculation, heroic possessions, and profits.

Distribution taxes. In order to reach this position, however, there may be a series of transitional taxes, such as an added tax on all new vehicles. Heroic income (over $1 million) taxes would be transitional until some form of equity would be established.

Under this new tax scheme, there would be no taxes on buildings, equipment or inventories. There would be no corporate income tax, although, as Daly and Cobb suggest, profits would have to be distributed to all shareholders as income. There would be no personal income tax, no property tax, and no sales tax. Why? Because things people want or work for should be encouraged. Once the real price of oil and other goods has settled out, no other taxes are needed for value-added things. These taxes could be collected proportionately at the community, regional, national, and international levels, to support programs and institutions, such as the UN.

Figure 208. Bible Law? (Credit: Google).

Experiment: *Is All of the Bible Relevant to Law?*

Some ministers and politicians are recommending Biblical Law to be part of common law, thus muddying the separation of state and church. At the same time, they are pushing for Moslems be rounded up, and imprisoned or exiled for trying to follow Sharia Law in the US, which they are not even doing.

What's the difference? Should it matter? The US is a secular state with laws that are separate from religious beliefs, all of which are tolerated in theory. Therefore, no citizens of the country can put their religious law above those made by elected officials representing all of those citizens. This is the reason this country has survived without internal religious wars. But now, these ministers and their pandering politicians are fomenting religious violence by trying to substitute religious laws for secular ones.

The pastor Steven Anderson is foaming about the death penalty for homosexuality, citing a verse from Leviticus (18:22). However, he overlooked other equally rigid requirements of God's Law, such as the right own slaves (Leviticus 25:44), to sell a son into slavery (Exodus 21:7), to kill a neighbor for working on the Sabbath (Exodus 35:2), or to kill your wife's best friend for wearing white shoes after Labor Day (Saks 5:0 Ave).

Other parts of the Bible are equally outdated. Men promoting their new religion wrote the bible at different times. While many parts of the bible still make sense, and always will, especially Jesus's words to love one another, even your enemies, many restrictions had to do with a fragile society being persecuted by Romans and trying to hold itself together.

These ministers and politicians are also using the Bible as an authority on abortion. The Bible was not concerned with abortion. In fact, in Greece, Turkey, Syria, and Israel at that time, people regularly induced abortions with herbal treatments to avoid too many children, which could not be fed or supported. Sometimes live babies and small children were put out on the hills to die from exposure. While this was done regretfully and painfully, it was done—because it was believed to be necessary to avoid later suffering and starvation. Abortion is legal and necessary today in many nations (including the US), although for different reasons, such as an unwanted child from rape, to save the mother, to avoid the suffering of a new-born, or to avoid the birth of a genetic or toxic monster.

On the other hand, *it is illegal and immoral to murder doctors and caregivers in medical clinics*, regardless of whether you believe that abortion is wrong. It is illegal and immoral to hunt down others with different beliefs and subject them to discrimination and violence. It is wrong! And those of you who do it are criminals, more so than those you persecute and harm. If the law offends you, change it. But do not pretend your moral stance puts you above the laws of the nation.

Experiment: *Ask the USA to Give Away Power & Leadership*

Imagining giving over all military power, except for local police or national guards, to the United Nations. Wars have been fought for resources and territory, as well as for religious and personal reasons, without an international referee that has power or respect. Many of our wasteful conflicts could be more easily resolved through a neutral power. Conflicts would still occur, but they would not be as dangerous or genocidal.

We have had wars for a variety of reasons, mostly related to territory or addiction. Wars are traps. Addictions, such as those to resources or oil or money, make it difficult to escape from a trap, a trap being a kind of energy well or gravity well. Of course, many cultures are addicted to the illusion of control and power. The U.S.A. is trapped in the belief that only it, among nations, can bring prosperity and peace to other nations, with trade or violence. Eventually the trap is escaped, or destroyed; it might collapse with its victims.

Let's stop being the bully among nations. We rarely admit it, but we are. We invade anyone who snubs us or threatens us impolitely. It might be educational to give up trying to lead. We could follow for a while. Ditto for Germany, China, Russia, England, and France. Let Switzerland lead, or Iran, Chile, or South Africa.

We could turn in, surrender, our 'weapons of mass destruction' to the UN—of course all nations would be required to disarm. We could relinquish our seat on the UN Security Council. We could withdraw all troops and weapons from all external military bases.

What would be the response? Some other nations might follow. Are we worried bout being invaded? We will have kept all nonnuclear weapons, vehicles and delivery systems; furthermore, many American households are prepared to rebuff invasions, as well as the searches by the IRS. It is just as likely that other nations fear that we will invade them (based on our behavior in the recent past).

Of course, we could continue the experiment by ensuring that every country have the same weapons, for instance, cruise missiles and rifles—some things, such as nuclear warheads and land mines are simply not selective enough to be useful or moral. The whole idea of fighting is to target someone you do not like, not just anyone who may be walking on the road or living downwind. The reduction in power and weapons could make conflict more personal.

In fact, personal conflicts could be used to decide stalemates by nations. Two representatives or 'warriors' would be chosen for a token contest by personal combat. The nation of the winner would be judged to be correct in its goals. Think of the prestige, think of the entertainment value. And, it would likely prove to be as effective as any way to determine right or wrong.

Part 16: Uncertainty, Conflict & The Future

Experiment: *Will We Have to Fight Resource Wars?*

Although many resources are distributed unequally over the globe, as a result of different kinds of historical geological processes, trade can allow access to those resources. However, as a result of long-term processes of inequity, from keeping people enslaved to cultural hoarding, many people have far less than others.

As a result of the unequal distribution of natural resources, including unincorporated waste and pollution, and the unequal distribution of materials and wealth between people, economic conflicts arise, often becoming violent political conflicts.

Population density was controlled by the traditional approaches to resources. In archaic societies, cooperation and consensus, as opposed to competition and individual exaltation, permitted planning to remain informal. Population growth triggered competition and conflict, which lead to positive feedback of the thing that caused the stress.

Now, some stress results from the misuse and overuse of resources, especially minerals and fossil fuels. Are we in an Age of Resource Wars? Or is it an economic war of buying and controlling oil and water. Will there be conflict over critical elements, such as iridium and palladium?

Just think of the implications for history books, and for the young being educated: Water wars, oil wars, rare metals wars, territory wars, reputation wars, religious difference wars. In a sense, all wars come down to a resource, whether metal, land or identity. I would not want to be a grade school student then.

The problems of conflicts over resources are responsible for significant losses, such as the loss of renewal, which reflects the inability of social systems to renew themselves or provide security and resources for their constituents. Any cool war is preferable to a hot conflict, which can destroy habitats and resources, as well as cause immense human suffering.

The use of resources by a people, where the replenishment rate is constant and the rate of use exceeds it, is a serial trap. This trap results in ecosystem degradation that is less reversible. The industrial age mistakes the rate of discovery for the rate of recovery.

Possibly, taxes could be used to control and share local and global resources. A severance tax, on items that are 'severed' from their context, would have the effect of limiting the use of nonrenewable resources, such as coal or oil, as well as slowly renewable resources, such as forests. A loss tax is a tax on losses from the capital base, that is, it is a tax on the destruction of resources, not just on their use or on their negative impacts.

These actions depend on a massive, cooperative effort. This effort can promise a greater equalization of opportunities to use resources that are becoming more rare and expensive, with openness and a joint sense of responsibility. As protectors of place, Nations have explicit responsibilities, from conserving ecosystems to managing resources and the distribution of rewards and power. Fighting is an expensive way to claim resources.

Experiment: *War Goes Away But Gun Violence Stays*

In the Nineteenth-century, armies were able to slaughter their opponents with advanced weaponry: Rapid-fire guns, higher caliber guns, bombs, and mines. After various wars, some people in some countries were able to get original or copies of the same guns. In the US, or instance, the Second Amendment protected the right to bear arms, to defend against invasion, although recent advancements have made the arms a different species, a new dimension for killing.

The essence of war is to defeat or destroy an enemy. Governments are efficient killing machines. On the average in the nineteenth century, states killed 3.7% of their own subjects. In the twentieth century, states killed 7.3% of the world population. The largest problem with war, obviously, is its scale. War is waged between nations and groups of nations against other nations or groups of nations. Even on a local scale, it often involves larger complexes of political links.

The good news is that wars have decreased. Most violence is now between individuals or between groups. The constant conflict of cultures and the loss of life, human and other, is tragic. But, this kind of tragedy results from a failure of cosmology; humans are responsible ultimately, not fate or chance. Humans all are equals, not subhuman others. Humans are tragic because they are responsible for their actions. They can choose a tragedy of the commons or of dictatorial control—or they can expand their cosmology to leave tragedy behind.

Maybe the problem is unlimited competition. Competition with other species and groups leads to conflicts, which grow in size and sophistication, gradually including noncombatants, crops, land, other species, and ecosystems, until the war is against large groups of human communities and then against nature in its various aspects from microorganisms to invasive plants. Conflict cannot be separated from other interactions, such as environmental destruction or inequities. Conflicts have been escalating, regardless of who won a cold war, or whether nuclear disarmament was achieved.

Nation states enlarged and consolidated their territoriality, which gave them increased capacity for marshaling resources. The nation states are closely related to large-scale violence, usually having to do with trying to consolidate their power. They usually have a monopoly on power. Except for individual guns for citizens.

Hunters prefer advanced technologies to locate prey and destroy it.

Many hunters carry automatic weapons. Criminals and police are engaged in mutual escalation, much the way predators and prey react to new advantages in the other. When criminals use cop-killer bullets then police get better vests. When criminals use nuclear bullets or rail guns ...

In the US, there was an average of two shootings per month, from 2013 to 2014, in K-12 schools. Shooters prefer soft targets. Six ideas: Don't name shooters, impose treatment on dangerous mentally-ill people, flag potentially dangerous students in schools, confiscate the guns of dangerous people, make gun manufacturers liable for military guns, and limit number of working guns that one person can own. More ideas are possible.

People who shoot up schools tend to use multiple weapons. Why would you buy ten revolvers or an automatic weapon unless you were going to use it inappropriately? Automatic weapons were designed to destroy people. When used against game animals or people, they tend to destroy the target completely. They certainly remove the element of sport.

Why do people attack schools more often than military bases or police stations? Why has no one ever attacked an NRA meeting? Is it because shooters tend to be target shooters rather than dynamic conflict shooters? Some nations ban automatic weapons for domestic use. One wonders if armed people are not just preparing for war and practicing on children.

Experiment: *Defund Some of the Overfunded Military*

The United States is the largest single seller of arms and munitions on the planet. Half of federal discretionary spending goes to defense. This is a psychological security blanket, since only relics are built. It is driven by the defense industries, for their own profit.

The defense industry is a virus that destroys healthy economies. The economy secretly becomes a permanent war economy (after Seymour Melman)—the largest activity of the government. Since WWII, half of all tax dollars have been spent on military operations, many frivolous or ineffective. We focus on jets, not safe cars or airliners. Money is thrown at research on weapons, but not technology for renewable energy. On stealth aircraft carriers, not reducing carbon that contributes to climate change.

What is the social cost of this focus? Bridges collapse, schools decay, and manufacturing is sent overseas. The language of power and fear masks these problems. At the same time, the military-industrial-educational-corporate establishment rewards corporations with free passes to their welfare state (but not for the disadvantaged or poor).

What would happen if we cut 95% from the open and secret military budgets? We would have to close a lot of overseas bases, but could keep state bases. We would still be protected by the National Guard, which was given extensive foreign war experience after 2003. We would still have the basic army and elite squads of troops. The Air Force, Navy and Marines would be reduced in numbers, but the Coast Guard would be strengthened. Prices for military goods, from bullets to toilet paper, would drop to near the real cost. And, we would be protected at the international level by the United Nations (a more empowered version, to be sure).

The other 95% could be used to fix every part of the national infrastructure, from highways and bridges to rail lines, public buildings and public monuments. There would be money left over to fund free education for every citizen and immigrant, to the college level. There would be money left over from that to fund complete social security for every citizen and immigrant, and to extend temporary safety nets for bankruptcy and unemployment, for foolish addictions, food stamps and catastrophic illness.

This approach would also have the effect of reducing drug and hospital costs to near real costs. And with the money still left, domestic farms and feed animals could be restored to health from the harmful methods of factory farming, forests could be managed skillfully, and damaged lands could be restored. In fact, more land could be added to the national park system, and to conservation and preservation areas. And, the money left from this would be put into savings to cover emergencies; some of it could be used to buy food and resources for storage for 7–10 years, again as a form of savings. We would be balanced and happy. Whether we would be attacked would not be as important as making the nation healthy and productive.

Experiment: *Questioning a Machine Image*

Industrial cultures desacralize nature. Since the advent of the machine image, the concept of the sacred has been reversed. In the primary view, the familiar was sacred. When modern cultures made the familiar trivial, it became profane. The quality of sacredness was bestowed on the unknown, on mountains, wilderness or sometimes children.

Science is mechanistic; the metaphor of mechanism by Descartes, that the world is a machine, implies that matter is inert, everything is replaceable, everything can be controlled, and that machines must have a creator, that is, they are not self-making. Science is reductionistic; anything can be taken apart and analyzed to be understood. And, science is single-visioned; one natural interaction, competition, is basic, and one interpretation is correct. Science is perfectible; anything is perfectible. These ideas are parallel to the idea of unlimited good, where anything, even virtue, can be multiplied indefinitely. The invalidity of these principles come with the recognition of limits.

Analytic science has reached its limits. Data and information developed by hard studies have undercut the paradigms that guided their investigation. Yet the view is still dominant and being used. Unfortunately, the language from a mechanical world view dominates even ecologists and politicians. This world view impoverishes humans by claiming all consciousness for humanity. It claims that nature offers no joy, or love, or peace, or certitude. Emphasis on the evil of nature creates a gap between humans and their universe. In contemporary cosmology, there is no room for the intrinsic worth of nature.

Discussing limits to growth, or any limits, is regarded by many people as a defeatism, as pessimistic, and a blow to human growth. Unseen limits have as real effects as seen limits. A community is forced to accept an upper limit, beyond which it cannot grow any further. This is related to the carrying capacity. Denying limits does not make them go away. Furthermore, as William Catton has pointed out, there is a difference between raising the limits of carrying capacity and simply permitting greater overshoot of the limits for as long as possible, with the threat of a greater and more catastrophic collapse later.

Industrial agriculture relies on large amounts of cheap fossil fuels to overcome limits of water, heat and nutrients. Industrial agriculture and capitalism have different logics and different goals. Many times capitalism forces environmental destruction, due to the limits of the system. The narrow economic gamble is to win big or lose big.

The industrial machine is out of balance and defective, but we have been making it run faster to process more material. The only things rarely considered are lethal variables, the end of oxygen, for instance. It is not safe to be limited by lethal variables, as Gregory Bateson recognized; closeness to limits reduces flexibility, that is, uncommitted potential for change.

How can we avoid possible disasters? We can use traditional societies as examples for some of our modern communities, with the understanding that the scale has changed, our technology is far more effective, and our cosmology has become more abstract and mechanical. For a modern community to prosper into the distant future, the community has to adapt a conscious ecological plan. It has to learn to mimic a mature ecosystem in terms of diversity, balance, flow and waste.

We can adopt an ecological perspective, which would fit economic costs and needs to the limits of ecosystems, and monitor the economic process would reduce wastes and pressures on natural processes. The coupling of agricultural productivity to a solar budget, and the conscious restoration of degraded systems, would contribute to the health of ecosystems. Communities should be designed for an optimal fit within the limits of the system. Ecological planning has to consider the limits of ecosystems, as well as the limits variability and stability, in deciding how human activities mesh with domestic or wild ecosystems. We should plan to keep our exploitation and population 70% below limits.

Experiment: *Do We Live in an Artificial Planet?*

Domestic landscapes have been simplified and manipulated for higher levels of use; fields and forests are managed by industrial methods and tend to have the characteristic of domesticated, controlled lands. Artificial landscapes have been almost completely modified and covered, for travel, urbanization, and industry, although a few wild species may be present. Two percent of the area of the planet is covered by completely artificial landscapes—industrial or urban (two for water). Our agricultural ecosystems cover a larger area, and these systems are controlled by our activities of plowing, using poisons and fertilizers, and removing large wild species that try to use it. In our control of artificial areas, which include many wild species, we could imitate the process of ecosystems by allowing birds, bats, and other animals opportunity to distribute seeds and energy to other areas or to access their prey, which may be our 'pests.'

But, could we extend our control to the entire planet and manage all the systems at once? Could we live in an artificial system on the scale of the planet? Biosphere II tried at a very limited scale and failed. Oxygen concentration could not be stabilized. Neither could nitrous oxide (although the occupants could have died laughing). How many plants and animals, bacteria and viruses are needed? Because of the size of biogeochemical cycles, most 'ecosystem services' cannot be duplicated on the human scale of technology, in a building or a city.

Still, the dryballs of academia present the world as a kind of awful domestic garden, devoid of wilderness, inhabited by useful good species and soon-to-be-gone useless ones that dare to live in a humanized suburbanity. While I am not guilty of floccinaucinchilipilification, I will offer a few charientisms on their vecordius ideas. Instead of a wild nature, these lexiphanic, nescient Anthropocenic addlepates decree from their plastic chairs in concrete-block buildings illuminated by computer monitors that, based on the shifting foundation of ignorance and error, humanity will preside over working biosystems controlled by online college-trained scientists and engineers. These nugatory sciolists confess to causing massive extinctions, but ask us to believe they can fit together pieces from alien species to manage the domestic garden skillfully and profitably. Oh, the horror! We must require them to pass an easy test to see if the planet really is domesticated: Can they direct tsunamis to gently water drought-stricken lands? Can they control hurricanes to power wind generators? Or, can they harness bacteria to stop infecting corn? No? Just bavardage after all.

Perhaps we could try to manage regions or whole ecosystems, using adaptive management techniques such as benign neglect, e.g., by lowering our impacts and relying on the natural capital that is in the system. Natural capital, especially on a global scale is little understood. We are not analyzing it, documenting it or monitoring it. As for reducing the population to the ecological limits of production, that could be calculated, for instance, by using Eugene Odum's method for the state of Georgia. But, how would we get people to move or have fewer children? Those things, even if successful would take time, up to a human generation or two.

Another strategy would be to become attuned to the earth, to commit our fate to nature, and not just say that we have faith in modern technology to save us with an artificial environment. We must be flexible, not detached or noncommittal. We must commit ourselves and be able to adjust to necessary changes—to be in a state of risk. We must maintain an environment for bacteria, plants, animals, and humans that is healthy for all. Conservation and restoration must be tempered with preservation. We are too ignorant to tamper with everything. More researching is necessary.

Flexibility is needed; within limits, a variable can move to achieve adaptation. An ecological plan must create flexibility and then prevent civilization from immediately expanding into it. Flexibility, remember, is uncommitted potentiality for change. Flexibility must be distributed among many variables in a system. Freedom and flexibility in regard to most variables is necessary during the process of learning and creating a new system by social change. There are still many possible futures for the earth and humanity, but they become fewer as we burn or destroy our options.

Although the nature of the biosphere is largely determined by evolution, and by wild organisms adapted to specific parameters and to each other, the anthroposphere tends to be artificial and managed, with only human needs considered. We need to keep as much of the natural world as possible in the anthroposphere, and keep a large wild planet outside the sphere of our domination. There is a human need for variety, individuality, and for understanding the nonhuman. Emersion in trees and bees is necessary to nourish human attributes that are in short supply: Awe, compassion, reflectiveness, and brotherhood.

Figure 216. Roses in Altazor garden.

Experiment: *Short Shallow Answers To Long Deep Questions*

One can weary at the vapidity of deep questions, as well as at the good intentions of long, bombastic replies. Short answers are possible and may be preferred. Here are a few:

- What, who are we? We are the descendants of humans.
- Where do we come from? Other humans. Parents.
- Where are we going? To find a way to live well and better– and support younger generations until they take over.
- What is the meaning of life? Change and surprise, tempered with manners and respect.
- What happens when we die? The parts get recycled, but some actions get recorded in media or as patterns in matter.
- What is love? Respect for another, and caring for someone while they find their own way.
- How should we respond to terrorism? Respect them and help them find their own way (after prison time if necessary).
- Do animals feel and think? Yes. And sometimes they care for you while you find your own way.

That's it folks. Can we move on? Philosophy aches to complexify our answers so we seem to be noble pure spirits, and maybe we could achieve that if we did not have to spend all our time repairing our errors in buying, killing species and simplifying the earth.

Science just cannot resist taking things apart, poking inside and explaining everything. It may have to kill plants, animals and people to understand them, but that just a small sacrifice to understand, for instance, that a diet of sugar makes animals sick and obese. And, there is a whole universe in need of understanding.

Society and culture in general provided people with an identity and a code of behaviors. It commanded fidelity and obedience. But the order allowed sacrifice, valor and meaning. What was the meaning for a Mayan farmer whose crops were dying from a fungus?

Homo sapiens coevolved with the biota in Africa for hundreds of thousands of years, but after their expansion and escape, they outraced any chance at coevolution with other wild communities. It has been a slaughter of forests and ocean shores, of plants and animals, who were replaced by domestic versions to feed the growing wise-wise populations.

Even other human species, from hobbits to Neanderthals and Denisovans, were extinguished in the race to singleness, bigness and success. E. O. Wilson asks us to imagine the moral and religious issues that their survival would have created for us. Their extinctions let us avoid a moral dilemma. Wherever humans appeared, biodiversity disappeared, returning to the paucity of 500 million years earlier.

The social strategies of individuals and groups include a calibrated mix of altruism, cooperation, competition, elimination, domination, deceit, reciprocity, and defection. These have been used consciously to make sure that the chosen family or group survived.

Humans have charged right through the great maze of evolution, tearing down walls and straightening corners, using our size, mobility, and reproductivity to impose our will and our designs on nature. We had the power, and therefore according to some, the right to do so. Presumably we also have the right to collapse and die if we overstep our bounds.

Figure 218. Hobbit skull with normal *Homo sapiens* skull.

Part 17: Global Governance & the United Nations

Experiment: *Understanding War First*

We need to task the United Nations with creating an international forum for each war (or large conflict). The UN would appoint 'arbiters' who would examine the extended history leading up to the conflict.

This would be open to the entire public on media, from radio and television to the web and 'the next new medium.'

A fair group of people would then weigh the reasons for the war. If they were deemed important enough, then war might be allowed, although with new rules to constrain it.

If it were not important enough, they would then award damages to one side, which the other side would have to work with the others to repair, making sure both groups had to work together.

If either side disagreed with the verdict, they could challenge the other side to a contest. The second side would decide what kind of contest, such as soccer or chess. After the contest was played, the winner would claim victory and receive, or not award, damages.

If that were not effective, then war would be possible, again. Again under constraints. These would limit damage to civilians, cities, farms, and wilderness areas, such that the wars would only be played out between small professional armies at an undisclosed, remote location, brimming with recording and communication devices for replays on the web or television. The winner would get bragging rights and royalties.

If this was not satisfactory, then it would be incumbent on us to admit that we have a death wish as a species, in which case we should satisfy it immediately to avoid collateral damage to natural species and wild ecosystems.

Experiment: *How Would Disarmament Work?*

The fight over guns is bizarre, compared to the lead of Australia, which took everyone's guns voluntarily. On a personal level, people could ask their national leaders to stop corporations from making and selling weapons. They could ask manufacturers to stop making toys of grotesque weapons. Someone has to heal the schizophrenic breach in our moral lives, between human welfare and destruction for profit. Our technology provides wonderful opportunities for communication. One can see the concern for concord (or peace) in Russia, Germany, China, and South Africa on television, as well as traveling through these countries. Most soldiers really do not want to fight, and a clear majority of civilians do not want to die.

On an international level, the United Nations could require large-scale disarmament of nuclear and excessive weapons. Furthermore, disarmament could be accomplished within a week. Taking this first step would add to the prestige of the country bold enough to do it. This is not a new idea. Earl Osborn, founder of the Institute for World Order, proposed the concept of sudden disarmament in response to the tedious phase-out envisioned by most plans in the 1960s. An agreement would not involve much negotiation. The United Nations could post a police force to disable all military ordinance (even hammers and welding equipment are effective on quiescent weapons). In fact, Osborn says that "rapid disarmament would not be difficult. A 1,000 planes each carrying 100 trained inspectors ... could distribute ... these men at all major centers in Russia and the United States within 24 hours." This police force could be staggered by nationality to avoid cheating. And, the same force could be used to settle international disputes. There is precedence for using unarmed peacekeeping units to mediate between hostile groups, in Cyprus, Kashmir, and Lebanon, among other places.

How could disarmament work? Humans are one of the most pacific of large animal species; they have a tremendous capacity for kindness and decency. Violence usually results from fear and ignorance. Violence is learned and so can be reeducated. Morals can change according to conditions, and not be defined for all time. A slow gigatrend of disarmament has already removed weapons from households and cities. The atavistic obsession with violence as a solution to simple problems is inimical to an ecological outlook. The image of society should change from competition and aggression to cooperation and mutual respect.

If we fear for our safety, we need only remember the success of nonviolence in India or the success of guerrilla actions in Southeast Asia and Central America. This alternative certainly could be regarded as utopian, but then the arms race is incredibly utopian, with its policy of complete destruction that no one really intends to implement. After such a war, there will be no place left on earth, and that is the original meaning of utopia, no-place. Working for disarmament is working towards eutopias, the making of good places.

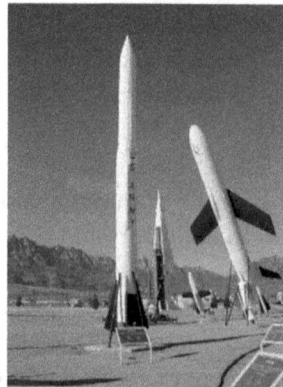

Figure 220. Disarmament targets
(accomplished in 2 weeks?)

Experiment: *Time for Global Emergency Designs*

There are serious emergencies. Species and ecological communities are dying out. Millions of humans are ill, unemployed or homeless. Industrial civilization seems to be on a path to collapse. In the case of industrial nations, we have been embracing excess for many generations, so that we are crippled by stress and sickness. Certainly there is an emergency if you consider the *extinction of species*, a thousand-times higher than background, or the premature deaths of millions of people every year. Certainly if we consider our civilized debt-loads and how little food and fuel reserves we have, or if we look at the natural devastations from natural events and poorly planned wars. It is an emergency. Animal and bird populations have decreased 50-90 percent in the past fifty years. What if that was the human population, the GNP or Walmart income? Would that be an emergency?

People are starving to death, possibly 50–60 million per year. Is that an emergency? Yes, and, it is a global emergency, caused by an emerging global system of industrial capitalism that is being adopted by most of the major cultures. Unfortunately the goals of the system are growth and profit, at the expense of the health of natural and human systems.

How should we act? Science has not been normative enough to say what we should do. Technology is too busy growing to illuminate any one good possible direction. Governments remain dedicated to preserving the status quo of greed and acquisition in unfair ways. Maybe design would work, if it could graduate from gadgets like chairs, toasters and swizzle sticks. Good design or ecological design could focus on whole projects, the entire psycho-socio-ecological network of civilization.

The nature of an emergency requires everyone to drop their normal activities and normal behaviors and to respond to a catastrophe. The catastrophe is usually quite evident, a wall of fire or a massive surge of water that will destroy or has already destroyed homes and people, as well as insects and birds, plants and animals, and their habitats. We seem reluctant to give the causes of these catastrophes the status of real emergencies, partly because the catastrophes seem like natural events, such as a warming trend, and partly because they are related to our industrial habits, which provide us with necessities, as well as with comforts and luxuries.

The planetary emergencies exists due to a series of slow, long, large, invisible catastrophes that are resulting from the normal wildness and uncertainty of planetary conditions, as well as the human modification of and interference with planetary cycles and diversity.

The problems may be regional and global, but they can be addressed on a local scale and human scale. A locality can have the authority, the power to take responsibility and make decisions, for global problems that impinge on the local. It may not solve the global problem, but it will affect it, especially if other local communities exercise their authority.

An ecological or global ecological design project could address every aspect of civilization, adding goals, resizing formal, corporate or cultural efforts to survive and prosper. Good design can make it easier to act on a local level, for instance, by providing human-powered transport for all countries. Pedal power can trump gasoline, even those that get 100 MPG. By offering simple, good, lower-carbon stoves. By combining functions to improve social use—Victor Papanek suggests putting washing machines (or Laundromats) in playgrounds, so mothers or father could socialize and work while watching their children. Papanek notes that the easiest way to save resources and energy, and cut waste, is to use less. Conserve. But, we can also produce designs for conserving sand surviving.

Are we doing enough? Only 5% of US plastics are recycled. But, plastic bags are still status quo. People cannot take an easy challenge and respond with an easy change. This is why it has to be an *emergency*! We have to take actions *now*! Coordinated, simultaneous, large-scale, drastic actions. Even a poor action would be preferred to sucking our thumbs in pretend security in our small comfort and competence zones. These zones resulted from lucky planetary and ecological accidents over a 10,000-year summer, but *this time is over now*. We have to be responsible for self-restraint and self-reliance, for not interfering in the developmental operation of the wild planet. We have to direct our representatives into global emergency actions, and participate ourselves.

Perhaps all we need is a diet. The essence of a diet is to restore one to health, by restricting unhealthy consumption or by changing patterns of behavior. As societies and cultures may also be guilty of this kind of behavior, so they need to put themselves on a diet. Archaic and agricultural nations have been strapped by historical inequities and unfair trading. Their challenge is to avoid simply repeating the same errors and consequences in the rush to acquire minimum standards and wealth. The solutions for all nations include trying new ways of balance for self-reliance, paying attention to cultural and physical catastrophes, and striving for better equity. Because of the extent of our overuse and misuse of ecosystems, and their effects on natural processes, systems are collapsing. As the catastrophes are large, the emergency responses have to be large, also.

We ought to reestablish energy patterns for regions, and use combined systems of wind, water, solar, organic and fossil fuels for energy. Singly, these may be inadequate, but as a mosaic they could meet decentralized needs for houses, businesses and some industries. We could also use small, independent, sealed nuclear power systems.

Let us make this a year of celebration, where we start no new products, no new house starts, until the old ones are renovated or resettled. No new profits for individuals or corporations. Perhaps we could put off having children for a year, when we would be more prepared.

Using less seems to be connected with economic health. In a changing world, with most of us fearing change, design needs flexibility with a synthetic approach. It needs to be integrated, comprehensive, and anticipatory, as R. B. Fuller and V. Papanek have urged. We have to direct our representatives into global emergency actions, and participate ourselves. This is another function of design, to change human behavior.

Experiment: *What About Global Climate Chaos?*

It is a serious global problem, but 'Global warming' is an unfortunate term, as it implies a gentle tanning. It also implies a comfortable, warm future where plants grow more and people need less heat from coal. Global 'burning' or 'suffocation' might raise more alarms. How we use language influences our priorities. I prefer 'climate chaos.' It sounds serious, but understandable. Chaos is the name for the maximum of individual orders, from which macro-orders are derived. Chaos is not an order of all orders, but it contains all orders. It is positive disorder. Chaos is identified with entropy in information theory.

Is global warming a problem for the planet? The planet has been much warmer before, during the Jurassic for instance, when there was no snow, ice packs or glaciers, and when dinosaurs were suppressing mammals and expressing heroic size to exploit their environments. Of course the pattern of continents was quite different then and less spread out.

Climate carries a lot of uncertainty. Every process on the planet contributes to the expression of climate. The planet has been hot and ice-free, as well as cold and ice-locked. When climate becomes uncomfortable for those living beings who fit it currently, it becomes a crisis. Climate change will alter the parameters of fitness for many species. Human culture will shelter humans from much of that, although the size of the human population may be a problem.

Warming now is pushing rainforest animals from the canopy to the understory, and from lowlands up the mountains, and from the equator towards the poles. Other problems associated with warming are ocean levels, and the nonsynchronous timing of plants, birds and insects. It will shift whole habitats polewards. It might get warm enough to push arctic specialists, like polar bears or penguins, out of existence.

The ocean and atmosphere are expanding. The ocean is going to swamp low-lying cities after a storm, or many storms. Over 50% of human dwellings are within 10 miles of the ocean. To save coastal cities, island cultures, and lowland agriculture, we might have to move inland. Warmer air is going to ruin corn and soybean crops. Bacteria and fungi become more virulent and cause disease and illness. Food and medical costs will spiral up. Animals and plants will get pushed to cooler areas, somewhere.

There are two general strategies respond to this chaos. The first is to employ proven or new technologies in projects for megageoengineering. Many of these technical approaches have to do with reflecting heat from the greenhouse atmosphere or removing carbon dioxide from the system by sequestering it and burying it for geological times. The low technical approach includes painting buildings white, switching to electric cars, using low-carbon concrete, and using snow-making machines to cover permafrost melts with white snow. A lot of carbon could be captured at the sources by current technology, and it is being done at a small scale. The high-tech methods start with thousands of airplanes spraying tons of sulfur dioxide at altitude for twenty years, although that is a temporary solution (and it might be effective). Another project is to launch thousands of mirrors in orbit to reflect light away from the planet. This project could create real problems if the mirrors move out of order.

The second strategy relies on ecological changes to fix carbon: Reforest as much land as possible; this would fix a goodly amount of carbon. Switch to organic, regenerative agriculture using perennial plants. Fit agriculture into urban structures. Reduce 'meat' farms, which takes 8 times the energy of vegetation and emits 51% of global greenhouse emissions; this requires a change in diet. Rewild the ocean and grassland systems; the animals would fix a significant amount of carbon. Repurpose corporations to build higher quality products and spend less on packaging. We could even have the Peace Corps and National Guard help people grow food locally, which would reduce some truck emissions.

The carbon dioxide in the atmosphere is going over 400 parts per million (PPM). This exceeds the threshold calculated by one group of scientists. Several groups, foremost is 350.org, are working to educate people and lawmakers to act to require industries to reduce emissions. Given the momentum of population and industry, it might be wise to aim for 240 instead of 350. So that we would reach a number *below* 350 as fast as possible, perhaps 275, the number after the Ice Age. Other greater threats may be methane, from permafrost melting, and ocean clathrates release.

Technological solutions might be fast, but they would have unforeseen complications and consequences. Ecological changes might be slow and insufficient, although they would point us towards a permanent state that would be healthy in the long term. Obviously, we should use both strategies, but assess and monitor the technical efforts, possibly rejecting some megageoengineering projects outright, such as orbiting mirrors. Changing so much, reducing so much will take fundamental changes in industry as well as agriculture. Large areas of the planet will have to be restored to more native conditions, although that should not affect cities, as much as roads and support lands (ghost acreage or footprints).

Experiment: *Can Failed Nations be Rebuilt?*

Cultures have been self sufficient for thousands of years. However, some of them failed. History records the debris of some civilizations that tried to manage their resources and failed; they existed in the Americas, the Middle East, Africa, Asia, the Pacific, and Europe. Some modern nations are experiencing difficulties.

Success as a nation can lead to hyperadaptivity and overshoot, to traps and failure. Hyperadaptivity is a serious condition that allows humans to adapt to poverty, bad diets, crowding, stress, suffering, and immense natural loss. Failure often follows tremendous success. The failure of nations in the past and present may be traced to the failures of cultures, communities and individuals. Failure can be related also to the properties of a culture, such as connectivity. Overconnection can be compared to power grid connections and failures. If overconnected, all can fail; if underconnected, many local areas fail; at a mid-range, transfers work.

The success of large nations may be due to accumulations of power, the wrong kinds of power. Power is the capacity to carry out reasoned intentions, even if the reasons are irrational. Power is centrifugal, as people have more influence over future. Power becomes a form of interference in the ecological support system. The cultures of industrial nations are often based on the unethical accumulations of power and materials.

Whenever there are too many competing for space and resources, whether genes, organelles, individuals, families, cultures, or species, some persist and some fail. Reasons for failures, such as poverty or pollution, start with a failure of imagination or will. Many people believe that energy and food increase automatically as people multiply, and that simplifying ecosystems can increase their productivity. This exemplifies the failure of imagination. We should not confuse the limits of our mind with the limits of the world, as the philosopher Schopenhauer warned. Failure to recognize limits is a failure of perception. The failures in our character or group or national character, can be seen to be responsible for the problems identified by Konrad Lorenz as the seven deadly sins of civilized humanity, from destruction of nature to the loss of civility. These problems can be described as a series of failures, from perception to intelligence and charity.

Nations are being linked in a partial global economy. It is important to rebuild failed nations to protect the integrity of the entire network of nations. For instance, the failure of neighboring economies can lead to the failure of trade, or the failure of negotiation. The failure of defense, the failure of contextual system, and the failure of a structure for individual participation can lead to the collapse in the meaning of participating. The fragmentation of social responsibility can led to an isolated self-image (unrelated to the external), to lack of self-confidence, and to a decline in participation. Conspiracies for societal control can lead to international conflict, government intervention, guerrilla warfare, and massacres.

The failure of a nation is a large political event, although its impact on the citizens of the nation may be minimal if they are self-sufficient. Nations can fail from single internal or external causes or from a combination of causes. These can include a weak cosmology, and image that does not fit environmental conditions, or poor knowledge, that is inability to recognize large and long-term processes, such as cycling. Internal conflict or economic collapse can diminish a culture or nation.

Sometimes a nation can rejuvenate itself with the passage of time (and natural processes restoring some damaged ecosystems). Sometimes a failing nation can be helped by neighboring cultures. To become healthy, a nation has to eliminate contradictory behaviors, especially regarding resources. It has to learn the maxima and minima of productivity and water cycles, so the population does not become too large to feed under adverse conditions, which do occur regularly.

Many practical things can be done. Most are matters of removal or restoration, according to an overall design for the nation. These things may start with removing weapons, bombs, and landmines. Then, we could remove obstacles to long-term success. Obstacles such as inequity, racial violence, and lack of opportunities to work and thrive for all citizens.

Some nations in danger of failing need external help. A first thing to attempt is the restoration of damaged or destroyed ecosystems. The restoration of forests from abandoned fields, anthropogenic deserts, and ruined ecosystems can be begun by something as simple as planting trees (and a good place to start is Haiti, which has little forest remaining; the US Peace Corps and other international volunteer organizations could start immediately). Of course, more complex actions, such as restoring the soil and reworking the landforms, might ensure that the restorations are more successful. Setting aside large wilderness areas, with the help of the UN or volunteer nations, would allow the self-maintenance of global cycles.

Richer nations have tried giving money, sometimes too much money. Most humans experience cognitive dissonance when gifts are too small or too large, as often happens. We could show some nations how to rebuild an infrastructure, such as schools and manufacturing plants, and even regenerative agricultural systems, and then help them do it, through international networks of volunteers. Temporary changes in trade would help nations achieve self-sufficiency first in food, then in shelters and meaningful work. Positive feedback of this kind could strengthen a weak nation or collapsing nation so that it could contribute to economic or cultural exchanges. With the increase in security, wealth, and self-esteem, cultures could be dependent on ecosystem productivities and still be diverse and unique.

Experiment: *Can an Extinct Culture be Restored?*

Usually when a culture collapsed, the people lost critical knowledge of it. However, often the people went back, when they could, to an earlier stage of complexity, e.g. swidden, herding or gathering. Sometimes they kept some of the culture, especially language and social rules.

Cultures could be rebuilt or restored, much like ecosystems, by applying general knowledge of their operation and components, especially those that are redundant and replaceable. Special attention would be given to the health or harmony of a culture. One approach would be to use properties, principles, standards, and practices to describe how.

Cultures have properties very similar to communities and ecosystems. One unavoidable property is participation in habitats and cycles. Another property is manipulation of natural systems. Every culture uses an economic filter to screen the environment for resources. Every culture addresses the equity of its members. And, every culture has to deal with interior or exterior conflicts and violence.

Cultures are based on principles that are derived from the interactions in a specific environment, although the principles may be more comprehensive than a single environment. The first principle may be stated as 'self-preservation;' every culture strives to survive the challenges of existence, and often prescribes behaviors in the self-interest of the culture itself, before the self-interest of one individual. This leads to another principle, that the culture should strive to be healthy, that is, to maintain optimum levels of vitality and interest. Each member has to contribute to the health of the community and to the health of surrounding natural communities. Members have to respect the principle of ethical behavior. Another principle is that members have to recognize the intrinsic value of other beings, as well as their connections and patterns. Members have to work to increase the self-realization of others through identification with them, the idea of the 'larger self ' as part of that principle. Finally, at least for this rough list of principles, members have to act responsibly as part of the community.

To restore an extinct culture, we would have to recreate its cosmology, and the principles, standards, and actions that flow from a healthy image. Cultures, unconsciously or intentionally, create standards to maintain themselves in a balance with their surroundings. One standard is to present healthy, inclusive, positive images of the planet and universe. Another standard is to address basic human equality, especially of opportunity or value. There is a similar standard for granting opportunities for other living beings; other animals and plants should not suffer, be interfered with, or be pushed to extinction as a result of untenable human desires or ignorance. Another standard is the adjustment of population or consumption impacts below the constraints of the ecosystem. Human populations do not seem to have any psychological limits on size. The limits have to be culturally explicit, and this is what can be made through standards.

One set of practices is related to cosmologies and metaphors; a culture has to provide and use appropriate images, as well as interpret them and restate them regularly, every generation, to reflect changing challenge and increasing knowledge. Another set of practices has to do with economic and social justice; a culture has to have ways to share wealth, to set limits on demands and needs, and to be perceived as acting fairly.

The language of a culture is usually dynamic, adding and dropping words and concepts with each generation. Where the language has withered totally, as with Cornish, or just been abandoned informally, as with Gaelic, it may need to be relearned and used regularly until the habit of the language is recovered. Through books, folklore, and computer databases, many of the particulars of archaic or traditional cultures have been saved and could be used for restoration.

Behaviors also may need to be sorted out in terms of effectiveness, or at least in terms of harm. Many traditional behaviors can be maintained or adjusted, giving the culture important continuity. Rules and taboos can be observed or introduced. Once people act together with language and common practices, they will be rebuilding a culture. It will be a new interpretation, of course, but in time, it should function as good as any.

Experiment: *Can We Save Civilization?*

Many problems of civilization are not global or regional or ubiquitous. Change can results from either random mutation or accidental behavior. But, conscious invention can produce new structures with new capabilities. If a global civilization is to form and to survive, it must develop into a completely new type of social system. A blind, consumer society cannot be transformed into a conserver society without radical constructive change.

Because modern civilization is dynamic and developing, it has the potential to change itself to be more stable and less intrusive. Individuals, groups, and nations have the ability to design social and living systems. We do not require more data or knowledge before we start redesigning systems. Buckminster Fuller said that to change something you have to build a model that makes the existing model obsolete—it is not necessary to fight existing reality. The pieces of the model have existed for hundreds of years; we only have to learn to put them together. We have the skills.

In this approach, a strategy needs to reduce the destructive impacts of civilization. Maybe we can design a general self-regulating feature to lessen environmental stresses when they become apparent. But, we need to be able to see and understand them first. Another thing that a new strategy can do is to 'dewesternize' civilization and let other cultures and nations contribute to a common vision. We also need to 'deglobalize' those things that should not be globalized, such as traditional cultures. Perhaps what we need is a form of cultural anarchy that could be coordinated through the United Nations or some new framework.

Culture is the process that gives meaning to the lives of people who live in the culture—it is not decoration or artifice, it is a body of knowledge that allows people to make sense of the world. We need an integrated global framework to 'dehomogenize' the world. We tend to be drifting toward a more homogeneous world since we are destroying local cultural options. Yet, if industrial culture wipes out all other cultures, there will be no rivals or successors. Traditional cultures are not failed attempts to be industrial; they are manifestations of the human experience of specific places. They allow us to be aware of our place in this area of the planet.

One serious question is what to do with those nonindustrial or nonurban cultures that choose to continue to be nonurban or nonindustrial, to develop in relative isolation. With their style of living, they could inhabit portions of wilderness areas. The boundaries would have to be flexible. However, we might consider limits on the technology that they could use. In general, boundaries are going to have to be flexible between nations, as many groups will choose to stay nomadic, and others will want to shift according to identity with other nations. The UN will provide an open framework to allow civilization, the sum of the genius of all nations, to develop.

Vaclav Havel has suggested that a fundamental shift is needed to change the direction of the current civilization. His deep conviction is that the only option is to enlarge the sphere of the spirit. It may not be enough to invent new machines and new institutions. The people of each nation have to contribute insights and to increase their consciousness of all realms of existence. We must develop a new understanding of existence on earth. Only by making a fundamental shift will we be able to create new models of behavior and new sets of values that can allow civilization to flourish.

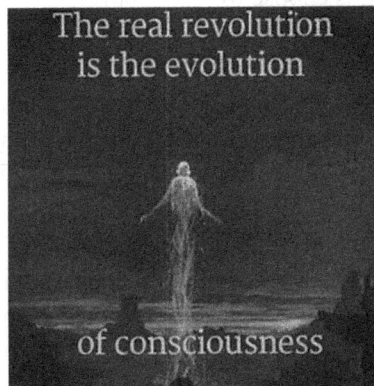

Figure 229. Saving Civilization?

Experiment: *Has Humanity Failed?*

Historically, we humans have used our skills and images to improve our lives and places. We have used our inventiveness to increase our ability to survive and to reproduce, as well as for our comfort and wealth. We have persevered despite many challenges and problems that have destroyed some whole cultures at times. We have survived ecosystem collapses and cultural collapses. We have survived droughts and freezes. We have survived diseases and wars.

We have used tools to changes places and some of the processes of nature. We have used technologies to protect our plants from insects and our towns from invasion. We have created the possibilities of unprecedented luxuries and wealth. We have made designs for stimulation or profit, more than for safety or elegance.

But, we have failed to understand much of the detail and scale of nature. We have failed to use our imaginations for things beyond luxury or entertainment. We have failed to share new wealth with most others or to share much of the planet with ultrahuman beings. We have failed to create designs that would incorporate us within the limits and cycles of ecological processes. We have failed to have the nerve or courage to try to limit our size and appetites, since it would inconvenience some or disrupt the wealth of a few.

We have the conceptual tools. We have religions that bind us to cultures and places—that teach us simplicity and charity. We have sciences that allow us to learn ever more about ourselves and nature. We have words and metaphors that can produce strong flexible images to guide our learning and decisions. We have the ability to create thought experiments and fundamental analyses of problems and catastrophic situations. We have the creativity to make ecological designs on a regional and global scale. We have the ability to expand our consciousness to include an ecological perspective of our situation.

But, many religions choose to teach hatred of other faiths, and ask that their own be enshrined by the governments of nations. Charity organizations become stealing contests for their leaders. Words and metaphors can be used to reinforce prejudices and reasons for fighting and extinguishing other people and species. Creativity can use its brilliance to confine us to abstract virtual worlds with 'trivial pursuits' to distract us from the problems of not enough food or shelter for many. Our consciousness chooses to expand inward for pleasure.

Furthermore, we have been seduced by our brilliance. We have become trapped by our close adaptations of agriculture and cities. We have become too comfortable in our habits and pleasures. We have allowed growth and inequity, dominance and slavery, to continue because it has become part of our perceived patterns of economic success. We have allowed conflict to continue because it is perceived as an unavoidable product of our success.

We are weak and arrogant, clever and clumsy. We ignore gigatrends and hide behind bad metaphors and the momentum of recent industrial growth—the financial size of that system was $55 trillion in 2004. We do not see the long-term, slow, invisible or other catastrophes such as extinctions and ecosystem collapses, although climate change is noticed enough to be explained away as not our fault. We neglect to question our adaptations in the past, even as they require prodigious amounts of energy that push basic ecological systems towards collapse. We chant our desired needs for luxury, money and growth. Then, we complain that the current local and global systems have no historical or cultural analogs to guide us. Everything seems inevitable.

Global environmental change is beyond our management or control, so business continues as usual, greenwashed or not. The massive destruction of ranges by livestock is an unavoidable product of economic growth, not a small by-product. People sliding into poverty and starvation are acceptable losses. The failure of ecosystems is less important than discovering further deposits of fossil fuels. The decline of two-thirds of planetary ecosystem support is forgotten in the excitement at the behavior of actors and politicians. Nothing makes us happy.

Virtually all of our actions are unsustainable in the short-term and long-term. Human civilization is living on borrowed time and stolen assets. The continuity of global interactions will lead to thresholds that when exceeded will trigger unwelcome surprises. So far, we have failed. We are not reforming or reapplying ourselves. And, unless we learn to, we will continue to fail.

Figure 231. Humanity Fails? (Credit: Google)

Part 18: Wake Up! It's An Emergency!!

Experiment: *Underestimating Change—Big Speed & Scale*
How many extinctions can planet take at once? The Permian event took
95% of all species; the planet survived and renewed life. In fact, the first
five extinction events were survived, although it took 5–10 million years for
richness and diversity to be reestablished.

We do not seem to be worried because we seem to underestimate
the consequences of our actions at regional and global scales. Now, we
are eating ancient sunlight and excreting carbon compounds into the
atmosphere, accelerating the interglacial warming of the planet. But, we
have not worried as much about water vapor, a greenhouse gas, and that
is increasing fast and it can bind more heat that carbon. And, we have not
worried about methane gas, which is a greenhouse gas that can bind more
heat than water vapor. It is increasing dramatically. We cannot blame it
all on cattle or domestic animals. And, when the permafrost starts to melt,
many more millions of tons of methane will enter the atmosphere.

Whatever we do to the atmosphere is eventually transferred to the
ocean, which holds the heat, which eventually melts ocean ice and expands
the water (heat does that); it also makes the ocean more acidic, which
causes the deaths of organisms that could fix carbon or other elements.
The heat in the ocean could likely (as it has been involved in one of the
big 5 Extinctions) melt the frozen mountains and plains of clathrates on
the ocean floor, thawing the methane hydrates and releasing the methane
to enter the atmosphere in eruptions. Now, we should be really worried.
Where is the sudden wisdom to come from the understanding of the
magnitude of risks?

We have always underestimated threats, also. We underestimated the
rate of global warming. We underestimate the rate of Arctic and Antarctic
sea ice melting. We underestimated land ice melting from mountain glaciers
in Africa to Antarctica and Greenland.

We underestimated the scale of the event. It seems too large to
understand; how could the entire atmosphere react—it is so big? How could
so many plants and animals go extinct? There are so many, in so many
places? Could our little transformations to agriculture kill so many? We
underestimated the speed of events. How could the atmosphere heat so fast
(well, 200 years)? How could a few inventions have changed so many things
so quickly? Global events are hyper-events, beyond our grasp.

The global scale is not a human scale. The global tempo is not
a human time. We need to acquire a larger perspective that can cope
with nonhuman scales of space and time. The earth may survive a sixth
extinction even, but humanity may be too rigid and unwieldy to survive.

Experiment: *Surviving the Human Big Bang Emergency*
Every civilization encounters a risk spectrum of some kind, which may push the civilization to find new sources of food or energy. The need to solve such risks pushes the system further from its original state, sometimes at increasing speeds. The conjunction of long-term risks can be called a crisis. Usually a crisis emerges from many crises, and happens to be the most visible, and it often is merely the end of some long-term process that escaped detection by the managers of a city or state. Thinking short-term, combined with delays in responding to a crisis, can lead to collapse.

In Mesopotamia, salinization was a relatively slow process that be-came a problem over the long-term, hundreds of years. Many processes, such as salinization, are long-term problems and do not become evident for several generations. They are also very difficult to reverse. For a society than needs surpluses to continue, there is little flexibility to change. The only way to avoid the problems was to let the land be fallow for long periods until the water table fell. This strategy was impossible due to food demands. Many cities and empires collapsed.

Many nations still rely on irrigation, despite obvious problems with evaporation, salinization and pollution. Pollution is a symptom of imbal-ance and improper resource utilization. A more serious problem is our lack of understanding of the extensive, long-term effects of pollution on the soil or on the atmosphere. More recent technologies have found ways to neutralize toxic wastes with chemical treatments or by burning. But, these efforts are not keeping up with new forms of pollution. There is minimal reorganization to deal with pollution.

We might be approaching some limit to population growth, which could lead to a drop in the complexity of civilization, partly because of the processes of globalization. The largest, most unimaginable part of our recent predicament is the scale of change in the past 200 years—the human '*big bang*,' that is, the explosive expansion of the human bodies in economic and political systems. Based on cheap, abundant energy, in unique and high concentrations, machine power increased, and the factory system increased efficiencies and productivities, buoyed by capitalism and nationalism, and justified by the ethos of a mechanical nature and progress. All this led to the process and worship of growth, so that through positive feedback and igno-rance (or greed), growth became automatic and uncritical. *Human numbers increased 6-fold in 200 years, but the size of the economy increased 10 times faster than population, over 60-fold; and energy use has blossomed 80-fold.* The past 50 years has created more change than the previous 500 generations. The explosion of our *big bang* has propelled us past anything any human being or culture (or planet) has ever experienced.

There are dangers from this big bang: Slow catastrophes, such as extinctions and conversions (ecosystem losses) and sudden surprises, such as dead zones in the ocean or holes in the ozone layer as a result of overfertilizing (or overenergizing or overpoisoning). Other dangers, such as human ignorance of change and uncertainty, make it hard to respond to very long-term changes (as opposed to short 2, 4 or 9-year limits).

Why are we not responding to the challenges or catastrophes? Is it because the catastrophes are effects of our civilization, and we do not recognize them? And, when we do recognize them, we think the same mindset and tools can correct them at some later point, when it is more convenient and politically feasible. The reality of two-year business horizons and four-year elections for temporary security tends to suppress the unpleasantries of long-term problems, such as human suffering or a chaotic planet with deadly storms, eruptions and earthquakes.

Experiment: *Wake Up! It's An Emergency!!*
It's an *Emergency! Wake up!!* We have tried talking and calculating and throwing billions of dollars at small problems. But, this is an emergency situation, requiring large-scale, multiple approaches, with new technologies, massive conservation, and micro-energy solutions (all of which require participation), but not using old, unconscious assumptions and design traditions. This is not quite working. So, we need to try something else.

We need to act on a global scale. We have acted on a large-scale before, in times of world wars. We were able to treat war as an emergency and to encourage or enforce remarkable changes, such as rationing or redefining jobs. We were able to take these actions without destroying our citizens or our cultures (or most of the planet).

Although nature and our human nature, are not enemies to be vanquished, the current situation has similarities to war. Massive changes threaten our lifestyles. Resources are removed from our reach by thoughtless or inefficient use. Growing insect and animal populations are attacking our food supplies. Species are being forced to extinction. Habitats are collapsing. Dangerous chemical wastes are accumulating. Ozone holes are growing, extreme climates pressing, and the entire planet seems to be wobbling. Changes in climate and ocean balance, as well as renewed diseases and infiltrated toxic chemicals, threaten our lives. And, it is happening everywhere, at once.

Although we want to respond with a warlike approach, we have been fooled by the fact that we cannot see an enemy—fooled by thinking there *is* an enemy. We have been misled by the slowness and subtlety of the penetration of our defenses. We have been betrayed by our own desire to continue our industrial dreaming at any cost. Some people have noticed changes and have been crying alarms, but they have not been listened to. Everybody needs to be awakened; everybody needs to participate, everybody needs to sacrifice and work towards some peaceful solutions.

The big problems seem insurmountable, and simple actions will not save our civilization from catastrophes. Part-time participation will not be enough to reverse the degradation of ecosystems, and partial business greening will not stop the unraveling of global cycles. What is needed is an immediate, peaceful, comprehensive approach to this situation—wise actions in a framework of intentional, reflective design.

We need to implement a global practical framework for allowing the creative anarchy of traditional-size cultures to be able to implement appropriate technology to deal with their resources and with other cultures through a revitalized and empowered international body that has the power of taxing global resources and properties for its own support, as well as the power to disarm and neutralize the unhealthy influences of large nations and corporations. The framework could limit human expansion to domestic and artificial areas, by specifying responsibilities and duties, while permitting the free operation of nature on the majority of the planet. It could save neopoetic areas and reserves wilderness. It could encourage respect for natural and cultural capital. It could repurpose the Peace Corps and the military (and the unemployed) to help rebuild a national infrastructure and to help grow food locally with a permanent agriculture. It could repurpose corporations to recharter for real public service (as they once did). It could recommend recognizing limits and planning for them using an ecological perspective and a metaphorical approach—it would be the framework for global ecological design.

Why would global ecological design work? Because life has over three billion years experience with changing and adapting, because human life and cultures have over 50,0000 years of practical experience adapting and making changes, and because humans are immensely adaptable—if they can adjust to poverty and suffering, they can adjust to a few good changes. Perhaps it is already too late—limits may have been passed and the catastrophes cannot all be reversed. We do not know, and may never know, but we can still act as if we were wise, as if doing the right thing makes a difference. And, we will have worked together to help others, to improve things and to make good places. If we act *now*, this month, this week, this day, this **minute**, then the changes might be more effective.

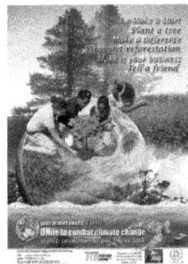

Figure 235. Plant a tree!
(Credit: UN).

Experiment: *Can We Change in Time?*

Our catastrophic situation requires immediate emergency actions on a global scale, from expanding personal consciousness to reforming the very character of human civilization and its coevolution with wild nature. First, an international organization such as the United Nations has to assume control of the planetary commons for limited use as well as for protection.

We need to end the thoughtless, large-scale experiments on ourselves, living communities and the planet, and imagine real thought experiments, radical ecological ones. We need to reduce our massive interference in ecosystems and landscapes, especially through conversion. We need to reduce and recollect many kinds of pollution, from plastic nurdles to carbon dioxide gas.

We need to rethink and rework design at every level, from personal and local to regional and global, through principles and actions, then use design to recover our tremendous losses, and to save and restore wild and common areas, especially extents of forests. We need to create programs to inventory places and the planet, then monitoring programs to keep informed as well as to reconnect many plant and animal patterns.

Human populations should be related to ecosystem productivity, as well as to rarity and diversity, even if this means a large reduction in numbers. Failed cities should be removed or refitted. New cities should be built with psychological and ecological knowledge. Technology could be reduced and integrated to fit needs and limits, instead of being automatic. Many predominantly technological problems, such as carbon production, fossil fuel extraction, nitrogen pumping, phosphorus loss, and pollution, could be solved through conservation. New smaller scale forms of technology can be designed using precautionary principles. Industry could be reformed as limited artificial ecosystems. Appropriate technology could be directed and certified.

We need to learn to be healthy individuals in living communities, to reduce antibiotic use and unnecessary luxuries. In fact, we need to place health in the context of ecosystems and global cycles. We need to learn to adjust to our own psychological and social limits. This means addressing problems with patterns of wealth, growth, inequality, poverty, and slavery. We need to link economics to ecologics, especially at the global level. We need to redefine the nature and limits of corporations, with new goals and community responsibilities. We have to consider the problems of intolerance, conflict and war within a frame of ecological ethics, and to reconsider the goals and responsibilities of nations and a global community. Failed nations and communities will have to be rebuilt, after restructuring international politics.

The design of new forms of human settlements and activities can be accomplished through ecological design. We can design the entire planet as a whole. We can integrate traditional adaptive patterns, from agriculture to technology and cities, with natural processes. We can strengthen religions with common understanding to bind people to their places and planet. We can make politics conscious and fair, with traditional goals and limits. We can make global designs, to restore the balances of systems, and to coevolve with the development of natural systems. We can create a planetary framework with clearly delineated functions at each level from personal to planet, in which everyone can participate.

Perhaps our efforts will be as naïve as continuous war, or as utopian as ecosystem destruction, but we have to try and to continue without pause or stop. Perhaps the ideas of global commons and small nations are as unrealistic as an infinitely growing economy or as warring for peace, but we have to try new patterns. We have to promote the appreciation of places, to awaken the delicacies and qualities of designs, to plan frameworks for development, and to allow everyone to participate in creating good, personal and social designs. Each single step is easy. Each challenge can lead to an adventure. Each change will have some effect. Each vision could inspire more coordination. And, all we have to do is to start to act, now!

Experiment: *We Can Find Joy in Play After All*

Humans are actors in a tremendous presentation. This metaphor once formed the basis of a worldview where nature was a theater of violent competition. The frame of the metaphor carried the conceptual baggage of the culture in which it originated. So, over time the play supported the idea of the superiority of "favoured" races in the struggle for existence and emphasized the role of competition in biological and cultural situations, at the expense of other "softer, weaker" interactions, like altruism or love.

Alas, the metaphor needs to be expanded. All species play a temporary role in the local stage in the ecological theater. But, all the actors and acts are essential. We are foolish to think some species as more important than others—that is ignorant or wishful thinking. All species and things contribute to the functioning of the whole, including rocks and gaseous elements. But, some are invisible to us because of their size or longevity. Some play their roles in a clump of soil, others in continental landscapes. Some acts last less than a second; others take millions of years. Even if actors seem to leave the stage or the act seems over, they influence the play with evidence of old impacts and with their corpses and elements.

There are many stages playing simultaneously in the theater, and it is not a one-act play. The human play has converted some of the stages, subverted others. The human actors are ridiculously egotistical and ignorant, pretending that the stages and theater is for them only, and that others are supporting characters. They pretend that the less important people are in the audience, but the only audience is on other stages with partial perspectives.

And, some plays are embedded in others. One play is the immense evolutionary play, where autopoetic beings drift through filters in the morphogenetic landscape. Another play is the maxed human conversion of ecosystems and the urbanization of human communities, with some pests and companions.

We have trouble understanding the theater or the plays because of our physical, temporal, psychological, and cultural limits. We see species and ecosystems as individuals, when most of them are in fact communities. We see walls and barriers, and not permeable filters. We see philosophical constructs of classes in isolated locales. We have problems dealing with motion, indeterminacy, ambiguity, and vagueness. We are seduced by logic and fallacies to believe that we can understand and control the play.

Although the scales involved in a planetary theater are linked by processes, in fact, the scale of the planet and its evolving life forms is significantly larger and longer. As important as human changes are to us, and to the ecosystems from which we emerged and on which we depend, for the planet those changes may only bump the global system to another stable state, well within the range of those from the past 2 billion years, but possibly not to our liking. We are the stewards only of ourselves and our companion forms, not of the planet.

On the other hand, Francis Bacon noted that we faced four major obstacles in understanding nature: The idols of the tribe, cave, marketplace, and theater. The perspective of the tribe is our inherent tendency to interpret and measure nature through our human senses, with their limits and scale (which can be applied with care). The idols of the cave refer to our personal peculiarities, pride or sacrifice, for instance (which we can accommodate). Those of the marketplace refer to errors from language and logic (which we are always expanding). Most of the errors of the theater have to do with the fallacies of our world views, even those based on good metaphors. This last obstacle may be the most difficult to understand and to overcome.

Where does joy come in? We can feel joy that things are always new and different, that there is enough to continue to provide and to inspire. We are actors in a process, but we are genealogical actors in ecological roles in the evolutionary play in the global theater. As long as we realize that we are not everything or the center of existence, we can continue to thrive and feel joy.

Experiment: *Can We Live in Harmony?*

How can we live in harmony if people cheat and steal as much as they can? We cannot. Perhaps if we worked against the horrific inequities to balance the distribution and use of resources of all kinds, that would help. The equal apportionment of 'resources' to all cooperating participants in the global commons might be possible. This reapportionment would be enhanced by the wisdom of harmony, that is represented by a neologism: Harmosophy.

The first definition of harmony (from a Greek word, *harmos*, meaning 'fitting') is: The combination of parts into a pleasing or orderly whole. To harmonize means to associate different things in a proportionate arrangement. Proportionate is the due proportion of two things having a reciprocal relationship or an equality in measure. If nations were defined as having an equality of measure, not necessarily size or force, then the harmony of nations could be promoted.

In Chinese medical tradition, the highest good is harmony, especially social harmony, or good relations. A good person is one who creates and maintains harmony. Perhaps this is the best working definition of health. Harmony in a society is the agreement of actions, feelings, ideas, and interests that results in peaceable relations. Harmony is mutual constraint plus a shared adaptive history. It is constraint that limits the scale of a society. Harmony requires understanding and then planning for adjustments. Harmony requires adequate ecological information and the infusion of its significance in human affairs. And, it requires the time to do this, with the practices of caution and reverence.

Harmony can contain discordant notes and themes, and weave them into a rhythm. Harmonious action could change the patterns of domination of cultures and individuals. Harmony has to include conversation with other species, and trust in nature as a generally beneficent system, our mother system and home. People have needs, animals and plants have needs, the site has constraints, and these things can be married into a good pattern. Harmony is a constraint of the whole system.

Design is the participation in the process of the ecosystem as a harmonious system, with mutually restrained conflicts and constrained influences. An ecological design is a form of reciprocally constrained construction, where the organisms, environment and designs are co-implicative, co-defining, and co-constructing. They all engage in a process of self-assembly, where the whole is the whole system.

Designs are limited by the real biological constraints of ecosystem processes and biogeochemical cycles. We must know the constraints in order to create a healthy design. Our actions, like conservation, are a means to an end, which is human fulfillment in harmony with nature. The design has to work within the constraints of the ecosystem. Rather than emphasize static equilibrium, the design should emphasize dynamic heterogeneity and learn to adjust to disturbance.

There is a musical analogy with the health and operation of ecosystems, and with the health and harmony of nations—health equals harmony. Furthermore, harmony is related to wholeness. David Bohm, in his theory of the implicate universe, proposes that health is a result of a harmonious interaction of all the analyzable parts that comprise the extricate order—cells, tissues, organs, the body—with the surrounding larger environment. Health is a quality that is grounded in the total order of the environment (or implicate order).

The essence of harmony is allowing all elements to have a voice or sound. That is one purpose of the United Nations, to allow each nation to have a voice. Harmony is more likely if no one nation can dominate the others all the time at every level. Peace could then be said to result from harmony.

The observation and study of balanced relationships in complex systems should allow us to recapture an experience of harmony (and an intimation of divine from scientific knowledge of processes). For Arne Naess, his philosophy (Ecosophy) is one of ecological harmony. Living together in harmony, which Ivan Illich calls Conviviality, is a necessary strategy for a successful civilization.

The alternative to the ecological insanity of industrial transformation is wisdom. Wisdom is the functioning of a mind that is respectful of its own boundary and processes, according to Gregory Bateson. Wisdom is recognition of and guidance by a knowledge of the total interactive system. Wisdom is also the disciplined use of the imagination with respect to alternatives, exercised at the right time and in the right measure. But we need practical wisdom.

And, if we do not have it, Jonas Salk urges us to behave "as if" we were wise, by using good sense, that is, common sense. Wisdom is a new kind of fitness. To survive, we must accommodate ourselves to the new conditions of a radically different life. Survival in this sense is not a win or lose proposition, just success at living.

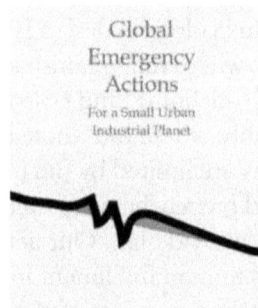

Figure 240. Book Cover

Experiment: *Taking Global Emergency Actions*

We need to be morally courageous. We need to initiate wild thinking, which has promise for our survival as an urban species. Sacrifice is called for, since we coasted along and waited too long. These actions could be taken simultaneously, since they fit together and should enhance each other. But, most of the actions need to be coordinated by some organization or global framework. We have to recognize that it is a global *emergency* and there are slow, large catastrophes. We need an agenda. Then we have to start to work, by implementing these and many other actions:

Mobilize! Increase grass-roots organizations. Pull political *leadership* along by showing how these changes are in their self-interest regardless of political philosophy

Give immediate *aid* to the dispossessed and starving, in the form of food assistance, education, and jobs

Start *sulfur dioxide* infusions to the atmosphere (up to 20 years); this is the only hi-tech solution to carbon fixation that might work—or not

Direct technology in appropriate ways for impact and scale

Reduce & capture sources of *greenhouse gases*, including direct-air capture

Start *carbon sequestration*, with biochar applications and industrial direct capture

Rewild the ocean immediately. Discontinue factory, large-scale fishing (Let nature do the work!)

Enhance populations of large ocean *animals*, whales, sharks, tuna, salmon et al (for their health and for carbon capture)

Empower the United Nations, or expand it. Invite *new nations* to vote in UN.

Disarm all nuclear weapons, as well as mass weapons. Close bases. *Reduce armies* to police forces.

Create larger UN 'peacekeeper' groups

Meet national *security* requirements with national guards, enhanced police forces, better wages, reasonable food production and prices, family planning and schools

Fit everything into education; transforming institutions; ecological thinking and the new paradigm. Create small community schools that can be funded with public school funds; infiltrate public and private schools by offering courses and presentations, as well as practical field experience and projects. Incorporate the organic dialectics of Goethe and the aesthetic education of Schiller; add in Novalis and Wordsworth in a deep ecological framework.

Start population *education* programs, with family planning, birth-control information, and financial incentives for small families. Start diet programs during this transition

Create *population* goals based on ecosystem food energy and on other factors, preferably using democratic means

Enlarge Peace Corps & *Volunteer* groups

Secure, protect and enhance *water* sources

Create hybrid *energy* network with energy capture (solar, wind, tidal ...
and small enclosed nuclear)

Start *closing coal* plants

Start ads and viral trends to *slow consumption*

Stop degrading wild nature. *Restore* degraded systems. *Replant* forests to
7000 BCE levels

Rewild savannas globally, using reduced cattle herds, then substitute bison,
elephants and ether grazers

Start *rationing* strategic goods, including water, oil, steel, and sugar

Switch to *permanent farming*, with perennial crop species (Land Institute),
Permaculture (B. Mollison), Natural Farming (M. Fukuoka), et al.

Switch to *ecological forestry*, with harvest goals & limits

Stop using fossil fuels, except for perhaps airplanes and some select heavy
industries

Stop dumping toxins in environment. Enforce *pollution* goals & controls

Shift to low-impact, certified *technology*.

Institute *steady-state economics*, with all-costs accounting (external and in);
distinguish between raw growth and favorable development.

Put *Taxes* on 'bads,' resource losses, heroic holdings and disproportionate
income ... but not on income, food or profits

Reinvent banks and corporations, with time and *activities limits*

Strengthen federal *governments* to issue money and expand social programs,
especially unemployment, disability and food assistance

Create millions of *jobs* for infrastructure rebuilding & city moving

Work towards increased global economic *equity*

Allocate & ration river waters based on ecological & hydrological limits

Change over industrial output to more rail, buses, bikes, heat exchangers

Survey natural species, resources and processes. Monitor them all

Designate wilderness areas and traditional cultural areas

Try to rebalance our individual lives; try to rebalance the social spins of
the patriarchal and matriarchal; balance the human and ambihuman

Try encourage individual self-reliance and health; try improve community
self-reliance and health; try to strengthen regional and global health

Strengthen community and cultural identity (against globalization)

Limit the centralization of power and authority, in style as well as trade

Reform corporations to act responsibly as public service organizations
(not imaginary individuals)

Try to emphasize environmental values, to repair the disjunction between
expression and practice, to accept the diversity of values in human
and nonhuman communities, and to reverse the lack of understand-

ing of connections or ethical models.

Accommodate competing views about the world, of what is important; relative conceptions of reality; vital needs versus necessary styles. Create a holocosmological framework to stop a prevailing industrial cosmology that is replacing place cosmologies

Address conflicts between human and ultrahuman, social justice and ecological sustainability, and resource consumption and policy.

There are more; this was just a short experiment. This coordinated approach may work in an emergency to avoid certain catastrophes. It is likely that these actions would all be eventually done anyway. The question is in the timing. Is it too late?

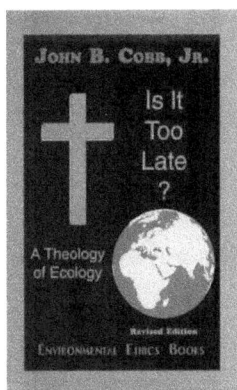

Figure 243. Book cover.

Experiment: *Is It Painful To Think?*

Arne Naess asked this question of ecologists and philosophers. We are asked to consider the long view when trying to "save" the planet. But, a long view seems meaningless when so much suffering or destruction already exists. Yet, an immediate, realistic, coordinated program of action is needed, capable of being implemented by communities and global agencies. We must face our responsibilities directly, declaring that there is no place for the excesses of corporations, governments, religious institutions, or technologies—we must act to end them. The declaration must be political, through cooperative networks or leaderless consensus, by persuasion and by example if not by fiat. The problem of human existence on the planet must be approached without deference to artificial boundaries of states, races, or castes. Poverty, pollution, and repression, as are extinctions and ecosystem collapses, are the concerns of every human community. We must stand and state that nature has limits, and that we all cannot have all that we want.

The application must be immediate. The crisis of exponential growth and destruction cannot be solved just after some final limit is approached or passed, or before a final exponential doubling. We must act without regard for imagined losses, financial or luxury. We know that whole countries have lost a generation and continued. We know they have rebuilt again from ruins. We could build from recycled materials alone, so there is nothing to fear from stopping. Immediate social reforms, the reallocation of resources, and the preservation of wilderness, are necessary, because of the nature of the problem; we cannot predict climatic or ecosystemic catastrophes. But, we have to start anticipating them now. Substantive change and research cannot be delayed until academic or political controversies are resolved.

We need to protect critical places in the planet immediately. Protecting my forest may not save global cycles, if continental forests are cut or if the ocean dies. Time is important. The losses continue to accelerate and grow. Species die; habitats unravel. We need immediate coordinated actions to fix economic unfairness. We should always do the right thing immediately, such as reduce resource consumption and driving. Then we must also work to change power structures to favor less damaging approaches. Also, radical actions can shift the political spectrum, much like ultraconservative actions. Then, the middle can be effective. We need to make hot triggers, with immediate feedback: Saving kilowatts, planting grasses or food, or lowering the thermostat. But, we need to include slow, long-term fixes, and do all of it together, because one or two steps will not be sufficient.

The transformation must be complete; it cannot be done partially. Local political and economic institutions have to change, as do regional and global ones. Holistic change will permit the reorientation and balance of local institutions. For example, air pollution is not independent of industrial processes, transportation, and employment patterns. Communities must be of a size that their members can feel responsible for them. These changes are demanded by new situations, ecological balance primarily among them. New institutions must be compatible with these new values.

The approach must be pragmatic and flexible. The readjustment to the realities of our new intricate involvement in the whole order of nature and her ecological balance will cause social strains. Some capital of energy and materials may be wasted. Population will have to be matched eventually to solar budgets or net ecosystem productivities, barring an incredible technological revolution allowing complete control at high populations, in which case we should emigrate off-planet. Production will be redirected to communal needs in transportation, housing, food, and recreation, as it is being shifted also to emergency actions.

The changes need to be coordinated globally. By its nature, a United Nations framework could reduce some of the stresses of transition, the uncertainty, ambivalence, or reversion. The UN could be a framework for local cultures and regions, where different human experiments are tried. Its variability would insure that we could reject any of the local visions that fail. This experiment may be called anti-human, anti-progress, anti-scientific, anti-technological, or anti-educational, but it is merely a new framework for conducting traditional human activities. A eutopian structure could offer movement towards common, achievable goals, starting with survival.

Taking these steps would solve many of the problems noted. The satisfaction of physical and cultural needs, as a result of living in stable and small societies, would contribute to the health of people. Fitting economic costs and needs to the limits of ecosystems and monitoring the economic process would reduce wastes and pressures on natural processes. The coupling of agricultural productivity to a solar budget, and the conscious restoration of degraded systems, would contribute to the health of ecosystems. Sufficient wilderness would allow the self-maintenance of global cycles. With the increase in security, wealth, and self-esteem, human populations could be dependent on ecosystem productivities and still be diverse and unique.

Human ills cannot be cured by a return to idyllic hunting and gathering groups or to a quasi-agricultural, ecologically caring society. There is no possibility of complete return. Most industrial nations are urban; agricultural countries have packed their surplus peoples in cities. Nor can there be a return to 4th century B.C. Greece, or to 17th century China, or to 1910 France, or to any time. Many traditional cultures no longer exist; others are disintegrating under pressure from industrial cultures. Nor can there be a jump to a complete technological future, where technology transforms hydrogen into wealth for everyone.

We should not underestimate the evolutionary potential of small cultures. We can afford to 'lose control' of other cultures, to allow their variability and experimentation. We do not need more information or rules, but we need meaningful ideas. Our attitudes and feelings toward nature need to be revitalized with evocative metaphors that let us accept responsibility for the part of the earth that we build, namely human culture and human landscapes.

Detailed designing and planning of complex open systems is not necessary. Designers or planners are not in a position to attempt detailed models of future situations because many relevant parameters remain unidentified, and many of those known cannot be quantified. Plans can be made within the limits of variables, although it is not safe to be limited by lethal variables, as Gregory Bateson warned. Closeness to limits reduces flexibility, that is, uncommitted potential for change. To minimize untested conclusions, this eutopian thought experiment is based on the values and forms of traditional cultures. This could allow rational planning to catch up, later. There is not enough time now.

To start, we need to know what is important and what is valuable; we need wisdom. We can start by listening to the pragmatic wisdom of the whole planetary system, with its billions of years of running itself. We can let it organize and evolve itself to create diversity and beauty.

We need to begin a prodigious new conversation. That which has been hitherto left unsaid—the goals of humanity, what we want to become, what we need to start to do now—could become explicit. Goals are not some final state reached once and for all time, but a horizon. It is time to define goals in terms of population, quality of life, and the preservation of biomes. It is time to identify and trace the catastrophes that have been forming and resolve how to act on them. A planetary electronic referendum could open communications, moderated by the UN and nations. A fast, web-linked social debate would let us address the complexity and levels of the emergencies; the UN and nations would coordinate the necessary actions. In operating in the planet, everyone can and should participate. This experiment offers continuity between goals, designs, and practical implementations, so that we can act now.

Experiment: *Act Now!*

A transformation of world order is necessary before the next large catastrophic event. But perhaps, as many have said, the dreadful has already happened—in Hiroshima, Vietnam, Africa, and lesser known places—and the cultural transformation is already awakening or has already finished. Most cultural transformations are invisible at first.

Maybe the gloomy forecasts are right. But, it is the function of a Cassandra to be always wrong. To be successful, Cassandra always has to be willing to be proved wrong. If she is disbelieved and her warnings go unheeded, then her prognostications may come true. But, if she is given credence, steps are taken and policies are changed to falsify her warnings. One suspects she would be satisfied at that.

Political response is not enough to change the system. Many problems are social or cultural; many are personal. As people change their attitudes, social and political changes can be made. This could take generations. A violent revolution could occur within weeks, but generations could be wasted imposing patterns on personalities. Social change can spark personal change.

The future is said to be in our hands, but our hands are in the cookie jar or on our genitals, and we are afraid to let go to touch something different. Perhaps we are just insecure. Disaster is not inevitable, unless we refuse to change our ways, in which case disaster is inevitable. So, we need to change. The change is going to discomfort many, but not as many who are being discomforted by not changing directions.

How long do we have before we have to change? A decade? A hundred years? When should we start? Last minute? Last week? Next month? How should we acquire urgency? Or the political will to change or the wisdom to change in better ways? What are our long-term interests? Is it necessary to declare a state of emergency? Another alternative? Not a war against nature or injustice, which is a tired metaphor, but just a rational response to a long, slow catastrophe. But, the response has to be immediate action.

After adopting an attitude towards catastrophe and starting to take action immediately, people, cities and a global framework will continue to act on those three levels. All three are necessary, but perhaps the individual is the most critical and important, if not to make changes, then to stimulate change for others and at higher levels.

This action has to include a survey of the state of the world, of all ecological and social systems. We have to make radical changes in the economic and political systems that arrange, produce and distribute wealth. We have to create a framework for world order without making the mistake of a gigantic global dictatorship, even if it is for the good of equality and environmental health. An order has to be based on bioregional models where nations are based on ethic populations in specific regions, where the operation is of interlocking hierarchical systems that make it easier for individuals to act. *We must act now!*

Figure 247. 50 Wildest Thinkers of the 22nd Century (after acting).

About the Author

Alan Wittbecker, observer, amateur and autochthonous forest-dweller, created a series of thought experiments on population and wilderness for *Pan Ecology* in the 1980s and for *Ecoforestry* in the 1990s. He wrote a newspaper column on Thought Experiments for the *Sofia Echo*, 2000-2002. Other experiments were posted on the *Thought Experiments* blogs (www. thoughtexperiments.us and www.syngeo.org and on *Facebook*) after 2004. No newspaper since has shown interest in such experiments as editorials, much less a column (*Sarasota Herald* and *Tampa Tribune* for example both refused to consider it). Over 100 of his articles have been published in international journals, including *Wolf Quarterly, Environmental Ethics, The Trumpeter, Common Good,* and *Wild Earth.* His books include *Eutopias: Making Good Places Ecologically and Culturally* (2nd ed, 1976), *RE: viewing thinking turning* (2002), *Good Forestry from Good Theories & Good Practices* (2003), *Global Emergency Actions* (2006), *Redesigning the Planet* (2007), and *Radical Ecological Thought Experiments* (2015). He is Director Emeritus of the Ecoforestry Institute, Senior Ecologist at SynGeo ArchiGraph, and he teaches environmental science and ecological design at Ringling College of Art + Design in Sarasota Florida.

Figure 248. Author (Credit: Lena Nieman)

Colophon
After the furious charge
Book composed outside Tallevast Florida
On an aging but fully operational Imac
tricked out with Pages, Indesign, Acrobat & Photoshop
Body text is Baskerville 11/14
Display Text is Gill Sans 14/14 & 36/40
Entertainment by Racer, a local ring-necked snake.

Figure 249. Never (Credit: Zazzle)